C083

Air Power

AS A COERCIVE
INSTRUMENT

Daniel L. Byman · Matthew C. Waxman · Eric Larson

Approved for public release; distribution unlimited

Project AIR FORCE · **RAND**

In Fiscal Year 1997, under the sponsorship of the Air Force Assistant Deputy Chief of Staff for Air and Space Operations and the Air Force Director of Strategic Planning, RAND's Project AIR FORCE began a two-year effort to explore the role of air power in future conflicts. The primary objective of this study was to help the U.S. Air Force (USAF) think about how best to employ air power to meet the evolving security challenges of the early 21st century. Particular emphasis was given to ensuring that air power would be relevant across the entire spectrum of crises and conflicts and that it would be effective against adversaries with diverse economies, cultures, political institutions, and military capabilities.

As part of this larger study, members of the research team explored the role of air power as a coercive instrument. In recent years, decisionmakers have called on the USAF to play a major role in attempting to coerce foes in the Persian Gulf, the Horn of Africa, and Europe. Although the United States and the USAF have scored some notable successes, the record is mixed. The purpose of the study reported here is to better understand the phenomenon of coercion and learn what is necessary to carry it out, anticipate likely constraints on the use of force, and determine how air power can best be used to coerce. The report will be of particular interest to USAF and other Defense Department planners who seek to use force more effectively.

The study was conducted as part of the Strategy and Doctrine program of RAND's Project AIR FORCE.

PROJECT AIR FORCE

Project AIR FORCE, a division of RAND, is the Air Force Federally Funded Research and Development Center (FFRDC) for studies and analysis. It provides the USAF with independent analysis of policy alternatives affecting the deployment, employment, combat readiness, and support of current and future air and space forces. Research is performed in four programs: Aerospace Force Development; Manpower, Personnel, and Training; Resource Managment; and Strategy and Doctrine.

CONTENTS

PART 2. SUCCESSFUL COERCIVE DIPLOMACY: LESSONS FROM THE PAST

PART 3. COERCIVE DIPLOMACY TODAY

Appendix

FIGURES

TABLES

The use of threatened force to induce an adversary to change its be-havior—coercion—is a critical function of the U.S. military. In the past several years, U.S. forces have fought in the Balkans, the Horn of Africa, and the Persian Gulf to compel recalcitrant regimes and war-lords to modify their actions. Yet despite its overwhelming military power, the United States often fails to coerce adversaries successfully or completely. In a number of recent crises, U.S. adversaries have openly defied the United States, complied incompletely with U.S. demands, or otherwise shrugged off threats of force. This report seeks to improve our understanding of coercive diplomacy, focusing particular attention on the contributions of air power.

The U.S. Air Force (USAF) will continue to play a crucial rule in future coercive operations. The USAF's speed of deployment and growing strike power make it a highly versatile instrument. In addition, the attributes of air power—including flexibility and precision—allow policymakers to alleviate constraints such as intolerance of casualties that often hamper coercive strategies.

CONDITIONS LEADING TO SUCCESSFUL COERCION

Several factors increase the likelihood that coercive diplomacy will succeed. First, coercive diplomacy is dramatically enhanced once a coercer achieves "escalation dominance." Coercion is more likely to succeed when the coercer can increase the level of costs it imposes *while denying the adversary opportunity to neutralize those costs or counterescalate.* When the coercer can manipulate the costs im-posed—and when the adversary cannot in turn impose costs on the coercer—the adversary is more likely to yield in the face of threats. Adversaries' responses to escalation, however, are varied and imaginative: Egypt in 1970 turned to the Soviet Union after Israel escalated to deep-penetration strikes near Egyptian cities; Chechen guerrillas terrorized civilian targets in Russia in response to Russian advances in Chechnya; and North Korea drained reservoirs to prevent flooding caused by U.S. air strikes against dams during the Korean War. Anticipating and minimizing the effectiveness of such countermoves are crucial to escalation dominance.

Second, coercion is more likely to succeed when the coercer negates an adversary's military strategy for victory. Such a "denial" strategy prevents the adversary from obtaining the anticipated benefits of aggression, making it more likely to concede. Serb leaders, for example, made concessions in 1995 because they recognized that they could not make further gains in the face of NATO air strikes—and might even be vulnerable to losses from Croat and Bosnian Muslim forces. Denial strategies are impossible, though, when an adversary's ambitions cannot be thwarted by military force or when the strategy it employs, such as guerrilla warfare, is difficult to counter even with overwhelming might.

A third factor contributing to successful coercion is the magnification of third-party threats. A promising coercive strategy harnesses the effects of threats against or military strikes on an adversary's vulnerability to another, external rival. Alternatively, threats or strikes can magnify an adversary's perceived worries of internal instability, thereby radically altering its anticipated costs and benefits of continued defiance. The primary danger of the latter strategy is that internal instability can operate unpredictably on regime decisionmaking, perhaps even hardening rather than weakening resistance to the coercer's demands. However, under the right circumstances this route has proven effective. For example, Israel's efforts to foment instability in Jordan forced the Jordanian government to crack down on Palestinian cross-border activity.

The historical record also reveals that several sets of challenges disrupt coercive strategies time and again. Perhaps the most important lesson is the need for accurate intelligence and careful estimations of the adversary. Failure often results because the coercer misunderstands the adversary's own perception of security needs. Another challenge is maintaining credibility. The credibility of threats is a product not only of skillful diplomacy in the immediate crisis but of past actions as well. Once the perceived will to carry out threats is undermined, even weak adversaries may feel confident defying the coercer, believing that any coercive campaign will be short-lived. Finally, some adversaries simply cannot concede. Some regimes fear their own downfall if they back down, while others are too weak to impose the demanded change even if they choose to yield in the face of threats.

THE CONTEXT OF COERCION TODAY

The lessons of past successes and failures of coercion must be understood within the context of U.S. foreign policy today and in the near future. Domestic politics inevitably shape coercive strategies, and the United States increasingly acts as part of a multinational coalition. In addition, the United States is increasingly likely to face nonstate adversaries, which pose distinct or heightened challenges to coercers.

The level and robustness of domestic support for coercive operations affect their conduct and chances of success. Air strikes are increasingly seen by the U.S. public and by many policymakers as a low-cost, low-commitment tool. Policymakers, however, regularly impose restrictions on U.S. military operations—such as restricting the size of forces, the type of missions, and the length of an operation—to gain public support. These restrictions may impede coercive diplomacy, preventing the United States from escalating or otherwise hindering U.S. efforts. Furthermore, adversaries can exploit U.S. domestic political concerns, particularly fears of U.S. and enemy civilian casualties, to counter coercive threats.

Another challenge comes from coalition partners. The United States regularly conducts coercive operations in partnership with allies, and coalition dynamics have tremendous effects on coercive diplomacy. Coalition partners offer the United States access to bases overseas, and their participation enhances domestic U.S. support for operations. Differences among coalition partners, however, restrict options for escalation. Divergent member objectives and responses often reduce the coalition's credibility, and adversaries can widen intracoalition rifts to further obstruct coercive operations.

The United States is increasingly involved in crises involving non-state actors, such as communal militias. Nonstate actors pose their own challenges, as they often lack identifiable and targetable assets. Moreover, the nonstate actor's very structure impedes coercion, because the nonstate group often exercises only limited control over its component parts. Although these problems can be circumvented to some degree by coercing the states sponsoring the groups in question, such a strategy multiplies the challenges listed above.

IMPLICATIONS FOR THE USAF

Used correctly and under the proper conditions, air power can play a major role in successful coercive diplomacy. The Gulf War revealed the awesome potential for modern U.S. air power to destroy a vast array of targets with speed and precision. This unparalleled capability, combined with the flexibility and versatility of air power, suit it for providing escalatory options, disrupting adversary military operations, or leaving an adversary vulnerable to a magnified third-party threat. As important as its destructive potential is air power's ability to restrict an adversary's escalatory or counter-coercive options. Finally, air power can help the United States avoid many common challenges to coercive success. By providing intelligence and helping observe compliance, air power can ensure that military force and diplomacy are effective.

Many of the constraints hindering the coercive use of air power are not technical—they are political and diplomatic. Air power's unique attributes may allow policymakers to alleviate public concerns about casualties, as well as those of coalition partners about extensive involvement, thereby eroding some of the key constraints that can severely degrade effectiveness.

However, the USAF must recognize the limits of military power: some adversaries cannot concede, and others will tailor their provocations to avoid being disrupted by air attacks. Given the public preference for air power over other types of force in situations where U.S. interests are limited, it may be used inappropriately, diminishing its credibility in future crises. By recognizing these challenges and limits, the United States and the USAF will be able to coerce more effectively in the future.

ACKNOWLEDGMENTS

The authors gratefully acknowledge the help of many RAND colleagues, who gave generously of their time and expertise. Alan Vick provided overall direction for this effort, made valuable suggestions, and otherwise played an instrumental role. Stephen Hosmer and Benjamin Lambeth also provided advice throughout the course of this effort, offering insights from their immense experience and knowledge about the use of air power. Zalmay Khalilzad and C. R. Neu shaped the direction of this effort with their comments and suggestions. Daryl Press served superbly as a sounding board and reviewer, freshening our thinking on coercion and helping us avoid many methodological errors. Thanks also go to Jennifer Casey and Gail Kouril for their research assistance and to Donna Boykin for her administrative assistance.

Professor Karl Mueller and Abram Shulsky both reviewed this document and greatly strengthened it. They offered excellent insights that shaped the organization of this work and sharpened the analysis, helping us avoid mistakes and simplifications. Colonel Hank Andrews also provided useful comments. A number of individuals at the School of Advanced Airpower Studies and the Air University also deserve our thanks. Colonel Rob Owen, Professor Dennis Drew, and Lt. Colonel Clay Chun all provided advice and a sympathetic audience to this project. Nora Bensahel, Kenneth Pollack, and Brent Sterling provided valuable comments on sections of this work.

Individuals in the Air Force Office of Strategic Planning played an important role in this project's inception and its development.

Thanks go to Dr. Clark Murdock, Lt. Colonel Julie Neumann, and Lt. Colonel Forrest Morgan for their assistance.

Any mistakes remain the authors alone.

Chapter One

INTRODUCTION

Coercion—the use of threatened force to induce an adversary to be-have differently than it otherwise would—offers considerable promise to mitigate, or even to solve, many security challenges facing the United States in the coming decades.[1] Yet coercion through military force rarely works as planned. Although U.S. military forces are without equal today, recent setbacks in Iraq, Bosnia, and elsewhere suggest that using this overwhelming force to shape even a relatively weak adversary's behavior is difficult.

Coercion is simple in concept but complex in practice. This study, which seeks to improve the practice of coercion, is organized around several fundamental questions—Why is coercion important? How does it work? Under what conditions does it succeed or fail? What is the context for coercive diplomacy today?—that develop the theory of coercion and show how it fits with practice. Answers to these basic questions will provide a backdrop for the larger focus of this study: developing principles to guide the coercive use of air power.[2] To this end, the study identifies the role air power has played in successful coercive operations, factors that have degraded its effective application, and ways that it might be used more successfully in the future.

---

[1]We use the term "coercer" to indicate the power issuing the threat of force and the term "adversary" to indicate the target of coercion. As discussed later in this report, even this distinction becomes muddy when an adversary tries to counter-coerce the coercer.

[2]We generally use the term "air power" rather than "aerospace power," because most of the examples, particularly in the historical section, involve air-breathing platforms.

COERCION AND U.S. NATIONAL SECURITY POLICY

To improve the U.S. ability to coerce adversaries, it is first necessary to understand the role coercive diplomacy plays in U.S. foreign policy. Recent political and geostrategic changes have elevated the practical necessity of effective but limited uses of force. In crises placing less-than-vital U.S. interests at stake, policymakers and the public alike usually prefer coercion over unrestrained, "brute force" solutions. Because many post–Cold War security threats pose at most an indirect or limited risk to vital U.S. national interests, the level of force used to respond will correspondingly be limited.

Geostrategic trends also raise the importance of coercion. The end of the Cold War has brought about the emergence of a world in which the United States has no peer competitor. Because of this lopsided force advantage (and, as noted below, the reduced potential for nuclear escalation), the United States has the option of using military force with little threat of a major defeat. In short, the chance of a coercive threat escalating into a full-fledged war should be more frightening to any aggressors than to the United States.

Accompanying the world of unipolarity and conventional war is increased uncertainty. Although the Cold War world was hardly as stable or as predictable as many people now recall, the U.S. military nevertheless had a well-defined mission: deterring a conflict with the Soviet Union and, if deterrence failed, defeating Soviet forces. Such a mission is thankfully missing today. Yet because the identity of potential threats is less clear, deterrence becomes harder while coercion becomes more necessary. Deterrence is more difficult when specific threats cannot be anticipated. The United States faces too many low-level threats to forecast and deter each one with a credible warning. The United States may, however, choose to react to threats after they materialize. At the outset of conflicts in Kuwait, Somalia, and Bosnia, U.S. policymakers did not issue a clear warning to deter aggressors, but later decided to intervene.

The end of superpower rivalry requires a corresponding shift in analytic emphasis. During the Cold War, the threat of nuclear war led to a focus on how nuclear weapons might be used to prevent, or limit, a broader conflagration. Success in brute force terms—reducing the Soviet Union to a smoking, radiating ruin—would be a failure in policy terms, even if the exchange ratio were 10:1 in favor of the United States. Today, the threat of total nuclear war is remote.[3] Thus, the context for coercive diplomacy that dominated the Cold War—two major, nuclear-armed powers locked in a zero-sum rivalry—is no longer the appropriate backdrop against which to consider the effectiveness of coercive instruments.

Nevertheless, one constant remains from the Cold War: military force is still a vital foreign policy tool. Sanctions, international law, and other mechanisms for affecting states' decisionmaking have proven neither reliable nor efficient in stopping aggression or abating undesirable behavior. Although military force may be the last instrument policymakers want to use, the absence of alternatives elevates the potential value of coercion.[4]

THE ROLE OF THE USAF

Perhaps more than any other service, the U.S. Air Force (USAF) will play a major role in future coercive operations and strategies. Several comparative advantages of air power make it a natural coercive device. The USAF is increasingly able to deploy rapidly and bring to bear quickly tremendous strike power around the globe. This speed and strength pose a potent threat to any adversary. Particularly when the United States seeks a quick resolution to a crisis, the speed of an air deployment will play an important part in successful coercion. Combinations of speed and lethality may enable the USAF to halt ground invasions or other limited aggression before a fait accompli occurs. Air power is also an attractive coercive tool because the amount of force employed can be discrete and limited, resulting in relatively few casualties on either side and enabling policymakers to exert considerable control over the scope and scale of operations.

---

[3]Because of the past emphasis on nuclear coercion, this work will focus on conventional coercion except when otherwise noted.

[4]For recent criticisms of the effectiveness of sanctions, see Haass (1997) and Pape (1997). For a critique of international regimes, see Mearsheimer (1994/1995). Kirshner (1997) offers a more nuanced account that describes different types of sanctions and their varying effects.

Technological advances, such as advanced sensors and communications, offer the hope that the United States can achieve its goals without massive force. Since World War I, air power experts have focused on the problem of finding, and then destroying, enemy targets. As Operation Desert Storm revealed, air strikes can now destroy a wide range of previously immune or impenetrable targets with relative ease. The improved accuracy of precision-guided munitions increases the potential destructiveness of even a small number of sorties.

The USAF's long-range strike capabilities are particularly useful for coercion. In contrast to the Army (or shorter-range carrier-based aircraft), the USAF can strike deep inside an adversary's territory, bypassing its conventional surface forces.[5] Ground power, on the other hand, requires first defeating an enemy's army before threatening an enemy's heartland. Long-range strike capabilities also make the USAF less dependent than in the past on facilities in allied territory. Finally, air power can not only strike quickly but can be withdrawn quickly; ground forces are hard to withdraw both during and following an operation.

USAF capabilities offer a potential solution to dilemmas resulting from casualty intolerance. Policymakers believe the U.S. public is increasingly unwilling to accept even small numbers of American casualties during military operations. Because ground combat, in general, involves greater risk of bloodshed than air operations, policymakers will often prefer air strikes over its alternatives when they expect that air power can accomplish the mission in question.[6] Technological advances may also enable the United States, particularly the USAF, to minimize enemy civilian casualties. During the Vietnam and Persian Gulf conflicts, U.S. leaders worried that enemy civilian casualties would erode American public support for the war

---

[5]This reach has long been the promise of air power advocates. Until recently, air power first had to defeat the air defense forces of the enemy, a process that itself often bogged down in attrition. Recent technological advances—most notably, stealth, precision guidance, and improved ability to suppress enemy air defenses—may place the United States in a unique position to avoid these difficulties. Future improvements in air defense, however, may again limit the ease of deep strikes.

[6]Eric Larson argues that policymakers misread casualty sensitivity during the Gulf War and that casualty sensitivity in fact depends on the perception of the stakes involved and the perceived prospects for success. See Larson (1996b).

effort. The ability of precision strike to reduce unsought casualties thus enhances the political feasibility of coercion.

Finally, the USAF offers a highly versatile coercive instrument. Air power can attack strategic, operational, and tactical targets. It can resupply friendly forces and provide essential intelligence. One, some, or all of these functions may play a role in successful coercion.[7] Future coercive strategies should be designed to harness these improved capabilities. This requires, first and foremost, an understanding of what factors enhance or impede coercive operations in general.

METHODOLOGY AND CASES EXAMINED

This study draws from a wide range of past attempts at coercion, including many that did not involve a significant role for air power. The cases were chosen using several criteria. First, high-profile and well-known cases were examined to ensure that the most historically significant cases, which are often the ones best researched by scholars, are included and properly understood. Second, the cases were chosen to show the limits and advantages of various coercive instruments and strategies—several cases were included specifically because they illustrate a rare, but important, point about coercion. Third, we looked at a range of coercing powers and geographic areas, thus reducing the likelihood of bias arising from the identity of the actor or arena which, in itself, should not infect the study of coercion. The cases in this study do not, however, represent either a universal set of coercion cases or even a representative sample thereof.[8] Appendix A lists these cases and briefly notes the most salient points for this study.[9]

The purpose of this study is to provide useful "rules of thumb" about the use of air power as a coercive instrument. It deliberately avoids a

---

[7] *Essays on Air and Space Power* (1997), p. 135.

[8] We recognize the methodological tension in building conclusions on such a limited sample. The conclusions we present should be considered hypotheses derived from the cases in question rather than theories tested on these cases.

[9] At the time of publication, the ultimate outcomes of Operations Desert Fox and Allied Force are still indeterminate. These operations are therefore not included in the appendixes.

narrow focus on whether air power can coerce by itself or other classic, but perhaps academic, themes of many studies on this subject. Instead, it tries to use the historical record to infer useful lessons about the proper use of air power and its limits. Several illuminating cases are deliberately given more weight than their historical importance where they suggest particularly valuable lessons for the USAF.

ORGANIZATION

The remainder of this study has four parts. Part One considers how to think about coercive diplomacy in general and argues that the traditional approach toward the study of coercion is of limited value to policymakers. Part Two surveys a range of historical cases to determine conditions under which coercion is more or less likely to succeed. With these general lessons in mind, Part Three examines the political and diplomatic context in which the United States will conduct coercive operations in the near future. It explores how coalitions and domestic politics will affect the ability of the United States to practice effective coercive diplomacy. This part also explores the special challenges associated with coercing nonstate actors. The fourth and final part considers the implications for the United States and the USAF, and it offers recommendations to guide coercive strategy.

PART 1. DEFINITIONS AND THEORY

To assess the effectiveness of air power as a coercive instrument, it is vital to clarify what coercion is and what it is not. Here we define coercion and suggest the best means to study it. The concepts introduced will be used both to survey the historical record on coercion and to examine the context in which it can be expected to be practiced in the coming years.

Our intention is to provide an analytic framework that captures the perspective of the warfighter. Among the most important aspects of this framework is its acknowledgment that coercion can at times backfire, leaving the coercing power worse off than when the coercive campaign began.

Chapter Two

HOW TO THINK ABOUT COERCION

Coercion is a commonly used term with no agreed-upon meaning. This chapter is theoretical, with the intention of providing a foundation for the rest of the study that includes the perspective of the warfighter.

We begin by defining coercion and noting the relationships among the term's many permutations. We then describe how to think better about this complex phenomenon, noting the value of a simple cost-benefit model and its many limitations. We then explore the limits in relevance and methodological problems common to previous studies of coercion.

Past studies of coercion that were based on unique geopolitical conditions are of limited relevance to the concerns of today's policymakers. Certain analytical limits also cast suspicion on their conclusions.[1] During the Vietnam War, defense officials explicitly used coercion theory in planning their strategy for the war,[2] and the same process is taking place today in regions that dominate newspaper headlines, such as the Persian Gulf and the Balkans. Analysts and scholars, however, have done little to advance our understanding of coercion, leading policymakers to repeat past mis-

---

[1]Similar problems plague the literature on deterrence. As Paul Huth and Bruce Russett complained, "Closer examination of many of the disputes, however, shows that much of the confusion is caused by methodological errors that produce conceptual muddles and inappropriate operational definitions of key concepts." Huth and Russett (1990), p. 466.

[2]Kaplan (1983), pp. 333–334.

takes or fail to learn from successes. As a prelude to our discussion of factors contributing to successful or failed coercion, we first provide a theoretical foundation.

DEFINITIONS

Coercion is the use of threatened force, including the limited use of actual force to back up the threat, to induce an adversary to behave differently than it otherwise would. Coercion is typically broken down into two subcategories: compellence and deterrence. Compellence involves attempts to reverse an action that has already occurred or to otherwise overturn the status quo, such as evicting an aggressor from territory it has just conquered or convincing a proliferating state to abandon its existing nuclear weapons programs. Deterrence, on the other hand, involves preventing an action that has not yet materialized from occurring in the first place. Deterrence would include dissuading an aggressor from trying to conquer a neighboring state or convincing a country that desires nuclear weapons not to seek them.[3]

Compellence, in practice, is difficult to distinguish from deterrence and to separate from the overall security environment. Such haziness often leads to misunderstandings of the inherent role that com-

---

[3]Among the most widely cited works from the 1960s and 1970s on coercion are those of Thomas Schelling and Alexander George & William E. Simons. See especially Schelling (1966) and George and Simons (1994). In *Arms and Influence,* Schelling developed the theoretical structure of coercion theory. He concluded by arguing for a strategy of gradually raising the costs of resistance, which could induce an adversary, eager to avoid future costs, to concede. In *The Limits of Coercive Diplomacy,* George and Simons expanded on Schelling's work, reviewing a series of case studies to draw lessons regarding the success and failure of coercive threats. Examining evidence from various international crises, this work analyzed contextual variables and other factors affecting the success of coercive strategies. From these events, the authors drew lessons of direct relevance to policymakers regarding the ingredients of successful strategies combining diplomatic efforts with minimal applications of force. The authors argued that a clear objective is necessary for coercion to succeed and that the precise terms of the settlement also must be specified. Moreover, assessing the strength of motivation—both that of the coercer and its adversary—is necessary. Also important is creating a sense of urgency, making the adversary realize that it cannot simply continue along the same path. Domestic support, assistance from allies, and strong leadership also are essential. Finally, coercers must recognize that perceptions are often more important than reality—the adversary must fear its costs, not just suffer them. (George and Simons [1994], pp. 280–288.)

pellence plays in deterrence and vice-versa. While analysts and academics typically draw sharp distinctions between the two, in practice deterrence and compellence tend to blur.[4]

Patrick Morgan's distinction between "general deterrence" and "immediate deterrence" provides a useful framework for dissecting these concepts. General deterrence involves preventing an action, whether it is planned or not; general deterrent threats are always present to some degree. Immediate deterrence focuses on a specific, planned event. An example of general deterrence is the U.S. treaty with Japan intended to secure Japan against any aggressor even though no state may currently menace Japan. An example of immediate deterrence is the 1970 Israeli warning to Syria not to invade Jordan: it prevented an imminent invasion threat from materializing.[5]

Reversing a completed action versus deterring a future, planned action (immediate deterrence) is rarely a clear-cut division, and both ultimately boil down to inducing the adversary to choose a policy other than that planned.[6] Classifying a case as compellence as opposed to immediate deterrence is always speculative to some degree, given the inherent opacity of enemy intentions. Indeed, even general deterrence and compellence are co-dependent, because the success or failure of coercion affects the coercing power's general reputation, and thus its overall ability to deter.[7]

---

[4]Some of these observations are elaborated in Schelling (1966), pp. 70–86.

[5]Morgan (1977).

[6]This distinction is far crisper in the nuclear, as opposed to the conventional, context.

[7]For works on the reputation effects of deterrence and coercion, see Hopf (1994), Huth (1997), Morgan (1985), Shimshoni (1988), Bar-Joseph (1998), and Lieberman (1995). Evidence for the reputation hypotheses is mixed (see Huth, pp. 92–93). In general, the reputation effect is stronger when it involves the same countries.

Figure 1 further illustrates the difficulty of drawing clear lines between compellence and deterrence. General statements such as "don't invade Kuwait" appear to fall clearly in the deterrence camp, whereas calls to withdraw would be compellence. The in-between areas are more ambiguous. "Don't go further" involves both stopping an existing action and avoiding a future one—both immediate deterrence and coercion. Moreover, a call to withdraw carries with it an implicit demand not to engage in the offense again and affects the credibility of the general deterrence call to not invade Kuwait in the future. These analytic categories have value, but the categories overlap considerably in practice.

The primary focus of this study is on the compellence subset of coercion, but given that immediate deterrence is a closely related phenomenon (both use the threat of force to manipulate an adversary's decisionmaking calculus), we incorporate insights and examples drawn from both subsets. We use the catch-all word "coercion" in the rest of this study. Unless otherwise noted, we exclude general deterrence cases from the study, although we recognize the phenomena cannot be analytically separated entirely.

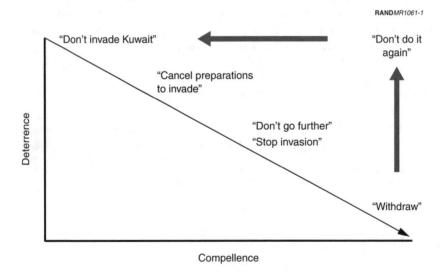

Figure 1—Deterrence and Compellence Blur in Practice

Coercion is not destruction. Although partially destroying an adversary's means of resistance may be necessary to increase the effect and credibility of coercive threats, coercion succeeds when the adversary gives in while it still has the power to resist. The Israeli demolition of Iraq's nuclear reactor at Osiraq destroyed the Iraqi ability to continue its nuclear program at that one facility. If that attack had persuaded the Iraqis to abandon their nuclear weapons programs altogether, the strike could be considered a case of successful coercion. But although destruction of the reactor achieved short-term results, it failed to change Iraqi policy and may have increased Baghdad's desire to acquire nuclear weapons.

Coercion can be understood in opposition to what Thomas Schelling termed "brute force." "[B]rute force succeeds when it is used, whereas the power to hurt is most successful when held in reserve. It is the threat of damage, or of more damage to come, that can make someone yield or comply. It is latent violence that can influence someone's choice . . . ."[8] Coercion may be thought of, then, as convincing the adversary to act a certain way via anything short of brute force. As Robert Pape argues, successful coercion is not war fighting; the target in question must still have the capacity for organized violence, but choose not to exercise it.[9]

Like the line between compellence and deterrence, that between brute force and coercion cannot always be discerned. Especially once an armed conflict begins, an adversary's behavior will always be dictated by a combination of both brute force and threatened (coercive) force. Pure or near-pure cases of coercion and brute force do exist. In 1994, the United States effectively coerced the military regime in Haiti to step down by threatening an imminent military invasion (Operation Restore Democracy)—no force was actually applied before the junta conceded (although the United States did send forces to help manage the transition) because the threat alone achieved the desired objective. On the other hand, Nazi Germany's 1941 invasion of Russia to conquer territory and seize resources (Operation Barbarossa) represents the brute force end of the spectrum—German forces conquered areas of western Russia without

---

[8]Schelling (1966), p. 3.
[9]Pape (1996), p. 13.

attempting to elicit surrender. Most often, however, cases fall along a continuum between these extremes.

Those who reject the brute force versus coercion distinction might respond that all state behavior, especially surrender, is always volitional to some extent. In no instance, the argument might go, has a state been so decimated in battle that it had a complete absence of choice. But there are degrees of choice that must be considered. Figure 2 illustrates that as an adversary absorbs more and more destruction, the proportion of the adversary's decisions that are motivated by the threat of future destruction declines. This is because the destruction of more and more of the adversary's assets narrows the range of options available to it and because, in some cases, the adversary has less and less to lose in the future. Brute force, by contrast, would eliminate the adversary's options completely. Operations against Nazi Germany illustrate this point. Even in May 1945, Nazi Germany was physically capable of continuing the war despite Berlin's capture. But it certainly had fewer options than it did earlier in the war at the commencement of the Combined Bomber Offensive when it had yet to suffer serious damage to its homeland.

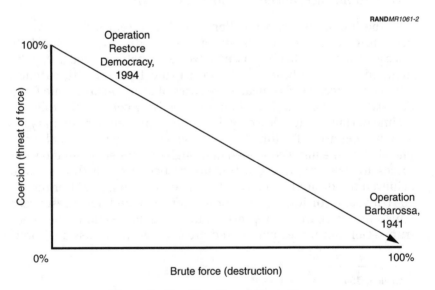

Figure 2—Coercion Versus Brute Force

From a policy standpoint, there is enormous difference between inducing surrender at the outset of an offensive and receiving it only after bringing the adversary to the brink of complete collapse. Indeed, this is of critical concern to today's warfighter. Coercion theory should capture this distinction by focusing on the use of *threatened* force to manipulate an adversary's choices.

A THEORETICAL STARTING POINT

Rational antagonists routinely use the threat of military force as an integral part of their diplomacy. Most standard explorations of coercion rely on an expected utility model to determine whether coercion succeeds or fails.[10] In his study of the effects of bombing population centers as an instrument of coercion, for example, Robert Pape uses such an expected utility model:

> Success or failure is decided by the target state's decision calculus with regard to costs and benefits. . . . When the benefits that would be lost by concessions and the probability of attaining these benefits by continued resistance are exceeded by the costs of resistance and the probability of suffering these costs, the target concedes.[11]

Coercion should work when the anticipated suffering associated with the threat exceeds the anticipated benefits gained by defiance.

The expected-utility approach to the study of coercion—critiqued in the following section—is a useful starting point for analysis. A close look at the cost-benefit model leads to a focus on four basic variables: benefits, costs, probabilities, and perceptions. Benefits are a simple concept, representing the value to the adversary of a particular action. For example, the benefit to Iraq of resisting U.S. demands to leave Kuwait was the possession of Kuwait and all that entailed. Costs—the other side of the coercion coin—represent the price an adversary expects to pay in pursuing a particular course. Perceived

[10]In addition to Schelling's work, a rationalist, cost-benefit approach is employed in other major works on coercion, including Bruce Bueno de Mesquita (1981) and Christopher H. Achen and Duncan Snidal (1989).

[11]Pape (1996), pp. 15–16.

costs and benefits include not only those associated with continued resistance but also those associated with complying with a coercer's demands. Probabilities affect both costs and benefits, indicating the likelihood that the benefits will be gained or costs suffered. Many actions could have high costs or benefits—Saddam Hussein might have suffered from a U.S. nuclear strike, but the probability of such an action is low. Other adversaries may perceive the same situation differently; their perceptions affect the relative weight decisionmakers give to various benefits, costs, and probabilities. Saddam, for example, might have understood U.S. intentions very differently than another adversary would have, and probably thought that the United States would not intervene to roll back his aggression in 1990. As discussed below, however, the standard cost-benefit equation is useful for heuristic purposes, but it has little value in helping the policymaker derive accurate predictions about a particular coercive policy.

Historically, coercive threats have been used to affect different points (often simultaneously) on the cost-benefit balance. "Punishment strategies" have tried to increase direct costs by threatening to inflict pain on an adversary's population or economy; "risk strategies" have focused on increasing the probability that the adversary will suffer costs by gradually ratcheting up the pain; and "denial strategies" have tried to lower the probability of benefits by making it less likely an adversary will achieve territorial or political goals.[12]

THINKING ABOUT COERCION: A POLICYMAKER'S PERSPECTIVE

Many coercion studies lack policy-informing lessons because of oversimplifications and mischaracterizations of the phenomena that coercion models are constructed to explain or test. Studies often lack relevance or face methodological limits. Because of these problems,

---

[12]Watts (1997/1998), pp. 129–130. Coercive strategies occasionally target an adversary's perceived benefits of resistance. The embargo on Iraqi oil sales, for example, reduced the value to Iraq of continuing to hold Kuwait (and exploiting Kuwait's wealth), and the freezing of the Haitian junta leadership's assets prior to Operation Restore Democracy reduced the value of continued rule.

the traditional coercion model on its own may yield little insight for policymakers.

Problem One: Limited Relevance

Many works on coercion have limited relevance to policy. First, the standard model's abstract nature often produces correspondingly abstract advice when used by academics to advise policymakers. Second, the model promotes uniform thinking, treating all adversaries as similar despite their unique attributes. Third, studies have focused on Cold War concerns, such as coercing a rival superpower, which are of little interest today. And, finally, many studies of coercion focus on coercion in war, ignoring the role that it can play in the entire spectrum of crises.

The standard model's elegance results in a lack of specificity when seeking to apply it to real-world problems.[13] Studies of coercion focus predominantly on the coercive instrument to see if it raised the costs of adversary resistance sufficiently to induce the desired behavior. From the expected utility model, one can take a case of failed coercion and reason that the strategy employed produced insufficient costs in the adversary's eyes to outweigh expected benefits. Yet this process tells us little. At times it approaches a tautology, as we learn only that coercion failed because the expected costs did not exceed the expected benefits.

At the same time, most studies of coercion ignore or overgeneralize the variety of adversary regime types and decisionmaking apparati. They employ instead a "one size fits all" approach to their conclusions, asserting that what failed for Iraq will fail in other cases even though Iraq's regime type is unusual. Indeed, the most basic model used by most coercion theorists—an expected utility model of decisionmaking—is acknowledged by many academics as deeply flawed and useful only when parsimony is valued more than accuracy.[14]

---

[13]The United States discovered this in Vietnam, when defense officials explicitly sought to apply Schelling's work to U.S. policy but found few practical ways to do so. Kaplan (1983), pp. 334–335.

[14]Zeev Maoz (1990) argues that in high-stress situations, individuals often face time pressures and make emotional decisions rather than ones based on careful analysis (pp. 318–321). Daniel Kahneman and Amos Tversky, basing their conclusions on

For analyses of coercion to be useful to the policymaker, they must take into account variation of key regime attributes. As Schelling noted:

> analogies with individuals are helpful; but they are counterproductive if they make us forget that a government does not reach a decision in the same way as an individual in a government. Collective decision depends on the internal politics and bureaucracy of government, on the chain of command and on the lines of communication, on party structures, on pressure groups, as well as on individual values and careers. (Schelling, 1966, p. 86)

Unfortunately, since Schelling's writing, analysts have made only limited progress on learning how regime variations shape coercive diplomacy. Simply taking a successful coercive strategy in one case and assuming that the same strategy will prove equally effective against a very different adversary is a recipe for disaster. Israel, for example, successfully used countervalue actions to force Jordan's King Hussein to crack down on Palestinian cross-border attacks from Jordan; similar tactics used in Lebanon in the early 1970s, however, led to the collapse of the Lebanese government and increased freedom of action for Palestinian guerrillas.[15] Yet the standard model of coercion does not help avoid this error.

In addition to problems with the standard cost-benefit equation, many existing studies of coercion fail to address the current needs of policymakers because they are colored by the particular conditions of the Cold War. These studies focused on coercing strong, nuclear-armed powers with powerful conventional forces; even when they focused on nonsuperpowers, they did so in the context of a world where the United States and the Soviet Union were locked into a zero-sum rivalry. The emphasis on nuclear coercion obscures the long-term implications of coercive strategies today. In a nuclear world, the legacy of an intervention is often a minor concern, one that pales in comparison with the benefits of averting a nuclear war.

---

empirical studies, argue that decisions in practice vary from the simple expected utility model presented above, that certain conditions affect how utility is judged, and that these conditions occur with regularity (Kahneman and Tversky, 1979). For a summary of relevant criticisms, see Green and Shapiro (1997).

[15]Shimshoni (1988), p. 226.

In a nonnuclear context, legacies and the effects of military actions on third parties become paramount. A massive intervention might sow the seeds of future hostility where more limited strikes might not. In addition, the limited use of force—not just the threat of force—is now far more plausible for most coercive crises than it was during the Cold War. The United States can often risk military strikes or more extensive involvement without triggering a nuclear-armed opponent. Thus, coercion involving conventional forces deserves additional attention.

A related limit of relevance in the existing coercion literature is its near-exclusive focus on major wars where the conflict is well under-way, not on small-scale conflicts or crises where hostilities have yet to break out. Because of this focus, most studies of coercion's effectiveness are unfairly biased toward a negative result. In the work of Robert Pape, for example, the "failures" of coercion against Germany and Japan during World War II or against the North Vietnamese are regularly cited as proof that coercion, particularly through bombing population centers, fails.[16] In such cases, however, the adversary had already devoted tremendous resources to the conflict, dramatically increasing its resistance to surrender. As Karl Mueller notes, Pape's work assumes that vital interests are at stake, but the implications are not clear in less-important crises where nonvital interests are at issue.[17] Smaller cases with lesser stakes would offer a more complete picture of coercion's effectiveness, yet they receive little scrutiny. This is especially unfortunate, as these small-scale conflicts or crises are far more likely in today's world than is a major power clash.

Problem Two: Measurement Pathologies

The study of coercion also comes up against formidable methodological problems. Thus it is not surprising that, despite the

---

[16]Robert Pape's work on coercion, for example, devotes most of its attention to cases that, in general, are more along the brute-force end of the spectrum such as World War II and Korea. Thus, Pape's work demonstrates that coercion through bombing population-related targets does not work when an all-out war is ongoing but says less about the importance of air power in coercive crises.

[17]Mueller (1998), pp. 192, 204.

impressive theoretical foundation built by past scholars such as Thomas Schelling and Alexander George, the study of coercion is not fully developed. Many studies do not capture the dynamic nature of coercion or wrongly treat it as a discrete event. Miscoding of data is also common—various studies have not consistently recognized the success or failure of coercion. These and related problems combine to lead analysts to misunderstand causality, which is ultimately where the military planner most needs insight.

The standard coercive "equation"—that successful coercion occurs when the expected costs to be inflicted by the coercer outweigh the adversary's perceived benefits in proceeding—is useful for understanding the problem of coercion in the abstract, but it may confuse the study of coercion when taken as a true depiction of state behavior. One problem is that this equation fosters static, one-sided thinking about coercive contests. It suggests that coercion occurs at a particular instant, at which time the perceived costs of resistance outweigh the benefits, leading to a change in behavior. It also encourages analysts to think about costs and benefits as independent variables that can be manipulated by the coercer, while the adversary stands idle and recalculates its perceived interests as threats are made and implemented.

A more complete picture requires viewing coercion as a dynamic, two-player (or more) contest. The adversary, too, can move to alter the perceived costs and benefits associated with certain actions. Since coercion does not, in fact, occur at a precise instant, the adversary may offset the costs associated with the threat. If behavior changes according to the balance of costs and benefits, then behavioral choices are governed not only by the coercer's threats but by the adversary's chosen responses. Before the commencement of coercion, at $T(0)$, the expected value of the adversary's behavior is presumably positive or else it would not so act. At $T(1)$, at which time a threat is made, the behavior will continue if the expected value remains positive. Even if the value becomes negative, the adversary may react in such a way that by $T(2)$ the value is again positive.

In addition to trying to minimize the effect of coercion, adversaries may try to impose costs on the coercing power. In effect, adversaries coerce the coercer. Adversaries attempt to force the coercer to stop implementing or reduce the coercive sanction applied by raising the political and military costs of implementation. Thus, at T(2), when the counter-coercion begins, the expected costs of continuing coercion (to the original coercer) may rise above the expected benefits of coercion success.

Nor does coercion have a discrete beginning. Coercion, or elements of it, occur all the time. Military capabilities, and the threat of their use, exert constant influence on all of our allies and adversaries, although in varying degrees. When we think about a "case" of coercion, then, we are in fact not talking about a sudden appearance of the threat of force. Instead, we are talking about relative changes in the threat of force—usually denoted by demonstrative uses of force, explicit threats and demands, and other overt signs. In other words, there is an ever-present baseline, or level of background threat, and we seek to examine deviations from, or spikes in, the level of threat.[18]

This dynamic approach yields three implications for assessing and designing coercive strategies. First, the likelihood of successful coercion is dependent not only on the expected impact of the coercer's decision but also on the adversary's countermoves. Second, coercive strategies must therefore be designed not only to raise the expected costs of resistance but also to anticipate and neutralize possible countermoves that would otherwise reduce the impact of the coercive threat or successfully coerce the coercer. Third, binary measures of success—measures that seek simple "yes" or "no" answers—are restrictive.[19] The existence of a coercive threat can never be measured in absolute terms; it is rather an increase in the relative level of force. To ask the question "Did air power coerce?" is misleading because a particular military instrument never operates in a vacuum. The true question is not whether air power "worked," but rather whether it helped or hindered coercion.[20] There is always a background level of coercion that itself fluctuates. The inde-

---

[18]These points are discussed in K. Mueller (1991), chapter 1; and in J. Mueller (1995), chapter 4.

[19]As Watts argues, mapping coercion to binary rankings is highly reductionist and wrongly assumes that complex campaigns can be reduced to zero or one. Watts (1997/1998), p. 136.

[20]Mueller (1998), p. 205, notes the importance of this contribution when understanding coercion.

pendent variable must be thought of as a marginal increase in threatened force, not the absolute level of force itself.[21]

The need to consider the background level of force against which a coercive threat takes place highlights an important difference between the analysis of coercion and the use of that analysis by policymakers. Questions like "Did the atomic attacks coerce Japan into surrendering?" or "Did the Christmas bombings coerce North Vietnam to negotiate terms more favorable to the United States?" represent vital, bottom-line insights for policymakers. Analytically, however, studies of coercion must be structured carefully to establish clear causal relationships between observed behavior and coercive threats. Using the Christmas bombings as an example, the proper question for the analyst to ask is "Did the marginal increase in force represented by the Christmas bombings lead North Vietnam to engage in behavior it would not otherwise choose?" An analysis of this question will yield the most instructive lessons for policymakers seeking answers in future crises.

The Uncertain Meaning of "Success"

In addition to the above problems in understanding the very nature of coercion, it is also difficult to measure the impact of coercive threats on an adversary. A danger when measuring coercion is to confuse instances of brute force, where the desired behavior is physically imposed on the adversary, with cases where the adversary retains the means to resist but chooses not to. For example, an analyst might reasonably code the Allied effort against Nazi Germany a success. After all, in the end German forces accepted unconditional surrender. Yet the sheer amount of force needed to force this surrender—and the near complete destruction of Germany's armed forces—would suggest that coercion failed: The Allies had to physically occupy the country to obtain their demands.

When analyzing coercion, analysts often forget that the threat of future force, not force itself, should cause the desired behavior in in-

---

[21]In addition, the time frame of success must be specified carefully. Some coercive attempts, such as the Israeli coercion of the Jordanian government, took years to succeed.

stances of successful coercion. Did the adversary choose to change its behavior, or was it effectively denied any option other than the behavior desired? After Israeli planes bombed the Iraqi nuclear reactor at Osiraq in 1981, Baghdad's nuclear program suffered a setback. The Iraqi leadership made no decision, however; the Israeli action simply removed Iraq's option of choice (the production of nuclear weapons from materials produced at Osiraq). The strike did not convince Iraq to abandon its nuclear weapons program and indeed may have stimulated it.[22] It therefore did not constitute successful coercion.

Another problem, alluded to earlier, is that many studies of coercion emphasize absolute, binary definitions of success rather than taking a more nuanced approach.[23] Studies of air power and coercion many times focus on such questions as whether air strikes alone compelled Japan to surrender during World War II rather than taking the more intuitive, and more analytically sound, approach of determining whether air strikes affected opponents' decisionmaking in a significant way. Even limited effects, when combined with other coercive instruments, may be enough to change adversary behavior. The current literature's narrow focus on whether a coercive instrument alone achieved objectives or failed outright leads to arbitrary and misleading coding of coercive strategies. Moreover, this focus emphasizes *whether* coercion worked, not *how*. Yet the *how* question offers the greater insight for designing future policies.

Classifying a case as success or failure depends on a precise definition of the behavior sought in that case. For example, in Operation Desert Storm the behavior sought from Saddam Hussein might have been his peacefully retreating from Kuwait. Or it might have simply been his not being in Kuwait, one way or another. One might conclude that the air campaign successfully coerced Iraq to withdraw

---

[22]Boudreau (1991).

[23]The use of these binary metrics of success stems largely from measurement concerns. If we wish to test certain hypotheses about coercion by correlating success with independent variables (such as type of force used or type of adversary assets threatened), then we would like to code as many cases as possible. A binary coding of success avoids the messy gray area into which many cases might fall if a nonabsolute measure were used. Moreover, clear successes are fairly easy to recognize: the adversary changed its behavior as the coercer desired.

from Kuwait because Iraq was willing to withdraw by the end of the air campaign under conditions relatively favorable to the United States.[24] Classifying the air campaign as successful coercion, however, assumes that the Coalition's objective was an Iraqi expulsion. But was that the objective? Janice Gross Stein concludes that the air campaign represented a failure of coercion because she has a different interpretation of what behavior the Coalition sought.[25] To Stein, the air campaign represented a failure of coercion the moment the ground war began, because Coalition objectives were to induce Iraq to withdraw *without having to forcefully expel it* through the use of ground troops.

The way in which the very issue of "success" is framed exacerbates this confusion. The use of absolute, binary terms—if a certain adversary behavior occurs, then coercion worked, whereas any other reaction constitutes failure—does not capture the dynamic and subtle effects of coercion. Iraq both conceded and defied the United States during Desert Storm. On the one hand, it offered a partial withdrawal from Kuwait as a result of the air campaign; on the other hand, it refused to accept all U.S. demands. The straitjacket of binary metrics distorts the lessons to be drawn from aggregated empirical data.[26]

As the Iraq experience suggests, states seldom respond to coercive threats with a clear "yes" or "no." When facing a coercive threat, states may modify their behavior, trying to placate the coercer with small changes while pursuing their own policy objectives. Syria, for example, sponsored terrorist strikes against Israel from Syrian territory after a new government took power in 1965. In response to subsequent defeat by Israel in the 1967 war and international pressure, Damascus no longer let terrorists operate directly out of Syrian territory. Damascus instead encouraged anti-Israeli terrorists to operate out of Lebanon, a nominally independent state that over time became a Syrian satrap. Capturing this modification of Syrian policy is difficult using a binary metric of success or failure. Yet the informa-

---

[24]Freedman and Karsh (1993), pp. 380–385.

[25]Stein (1992).

[26]For an example of binary coding of success or failure, see Peterson (1986).

tion is of critical concern to anyone wishing to understand the phenomenon of coercion.

Binary coding can also lead to the understatement of coercion's effectiveness because it neglects partial successes. In some instances, a coercive strategy—be it sanctions, bombing industrial centers, or another form of pressure—will move an adversary toward the coercing power's goals but not result in total success. During the War of Attrition between Israel and Egypt in 1969–1970, Israeli strikes on Egypt caused Cairo to abandon its goal of forcing Israel to relinquish the Sinai and changed Egyptian strategy from offense to defense—clear Israeli victories. Yet the strikes did not stop the occasional shelling of Israeli positions along the Bar Lev line or Egyptian sponsorship of Palestinian *fedayeen* attacks—obvious failures for Israel. Indeed, the War of Attrition prompted the deployment of an extensive Egyptian surface-to-air missile (SAM) network near the canal, causing Israel to lose local air supremacy—a loss that had disastrous results for Israel in its 1973 war with Egypt.[27] Such a limited success (or partial failure) is impossible to capture with a binary metric.

At the same time as binary metrics may bias studies against finding coercion successful, they may also understate the detrimental effects of coercive strategies. One of the greatest risks of coercion is its potential for backfire—threatening an adversary may provoke an increase in adverse behavior rather than the desired behavior. In 1941, the U.S. oil embargo and diplomatic demands on Japan led Tokyo to fear that the United States was implacably opposed to Japan's influence and that Washington would make further demands if Tokyo conceded. Thus, Japan's leaders saw war as inevitable, paving the way to Pearl Harbor.[28] The 1967 Arab-Israeli War and the 1969–1970 Israeli-Egyptian War of Attrition are frequently cited examples of inadvertent escalation resulting from coercive threats.[29] In Somalia, U.S. Army helicopter strikes on General Mohammed Farah Aideed's subordinates not only failed to intimidate Aideed but may have provoked anti-U.S. sentiment, contributing to the demise of the American-led operation. In other words, coercive strategies can

---

[27]Shimshoni (1988), pp. 125–186.

[28]Sagan (1994), pp. 84–85.

[29]See Stein (1991) and Bar-Siman-Tov (1991).

leave the coercer worse off than before. Within the binary framework, the worst outcome recognized is the null result: backfires and hardening of adversary resistance are coded just as if coercive threats caused no effect.

Conceptually, the dependent variable should be understood as a marginal change in probability of behavior. Against a fluctuating background level of threat (and blandishments, for that matter), the probability of the adversary altering its behavior, even without specific coercive action, is never zero. In other words, there is always a positive probability that the adversary will do the coercer's bidding on its own (as well as a positive probability that coercion will backfire and the adversary will become more resistant than before). Viewing success in absolute terms, based on observed behavior, ignores this positive probability and classifies all desired behavior as "successful" coercion, regardless of how likely that behavior was prior to the additional coercive threat. Although data limits may require a focus on observable behavior, analysts should not forget that the true effects of coercion can be found in the changes in the thoughts and motives of the actors involved.

CONCLUSIONS

This chapter presented a way to think about coercion. Like most studies of coercion, it relies on the cost-benefit equation as a heuristic device to focus attention on the roles of costs, benefits, probabilities, and perceptions. Yet because of the limits in relevance and methodological problems that come from too tight a fit to the cost-benefit approach, the study treats coercion as a more dynamic and complex phenomenon. There is no easy methodological fix. Rather, studies of coercion must avoid oversimplification and instead search for deeper historical understanding, even when (or especially when) the lessons are ambiguous. Successful future coercion will depend on understanding time horizons, recognizing unanticipated consequences, and otherwise incorporating context. The remainder of this study attempts to recognize these issues when seeking to explain the past record of coercive diplomacy and prospects for the future.

PART 2. SUCCESSFUL COERCIVE DIPLOMACY: LESSONS FROM THE PAST

Much of what is involved in coercion—inflicting punishment, threatening benefits, and shaping perceptions—remains constant. The historical record is not a perfect map for future coercion, but it does offer valuable insights into human nature and national decisionmaking that remain consistent over time. This part of the study examines instances of coercive diplomacy in the last 50 years, seeking broad insights into how to maximize the likelihood of successful coercion.

Chapter Three

EXPLAINING SUCCESS OR FAILURE:
THE HISTORICAL RECORD

Understanding the coercive use of air power demands insight into coercive diplomacy in general and the strengths and weaknesses of air power in particular. The successful coercive use of air power requires favorable conditions and often depends more on the strategy chosen by the adversary than on the overall might of the coercer. This chapter attempts to identify the conditions conducive to successful coercive diplomacy, paying particular attention to the potential contributions of air power. It also notes common challenges that may lead to failure.

The analysis reveals that the success of coercive operations is often a product of one or more of the following three factors:

- Escalation dominance

- Threatening to defeat an adversary's military strategy

- Magnifying third-party threats.

Coercion is most likely to succeed when several of these factors are present. At the same time, the analysis revealed that challenges in:

- intelligence,

- credibility, and

- feasibility

commonly undermine coercive strategies.

Successful coercive strategies not only build upon the three success factors but effectively avoid common pitfalls contained in the three categories of challenges. Appendixes B and C list the cases examined in this study and note the presence or absence of the above three conditions that contribute to success and of the three challenges that undermine coercive strategies.

CONDITIONS FOR SUCCESSFUL COERCION

Successful coercion—like success in any other complex phenomenon—can never be guaranteed. History is replete with examples of leaders and nations being blind to reality, fighting on despite every objective factor pointing toward surrender. Coercive diplomacy often is abandoned by the coercer or degenerates into brute force when the adversary refuses to concede. Nevertheless, the historical record does reveal that the presence or absence of certain conditions makes success more likely. In particular, dominance in escalation, ability to counter an adversary's military strategy, and magnifying a third-party threat to an adversary make coercive success more likely.

Table 1 describes Operation Deliberate Force (1995) and notes the relative presence or absence of the above conditions. The conditions for success and possible challenges are most realistically depicted as a spectrum. Intelligence, for example, is never perfect, but also is seldom completely absent. Similarly, escalation dominance is rarely absolute (most adversaries have at least some ability to escalate through terrorism or other low-tech means). A spectrum approach describes the degree to which conditions are present rather than simply their presence or absence. Such an approach provides a better perspective for judging the conditions for coercive success or failure.

Escalation Dominance

The ability to escalate credibly against the adversary—that is, to threaten imposition of a greater and greater price of defiance—allows a coercer to manipulate the level of costs the adversary associates with particular behavior. The cases examined reveal that

Table 1

Operation Deliberate Force: Coercive Conditions in a Spectrum

Condition	Positive <——————————————————————> Negative		
Escalation dominance	NATO dominant	XX	Serbs dominant
Threaten military strategy	NATO can stop Serbs	XX	NATO cannot stop Serbs
Magnify third-party threats	Croat/Muslim threat increased	XX	No third party present
Intelligence challenges	Good intelligence and estimates	XX	Weak intelligence and estimates
Credibility challenges	High NATO credibility	XX	Low credibility
Feasibility challenges	Serbs willing and able to implement	XX	Serbs unlikely to implement

the capacity to escalate is perhaps the most common factor in successful coercive operations.

But, as noted in Chapter Two, coercive contests are dynamic encounters, where *both* sides attempt to manipulate the other's cost-benefit calculations. Even if party A can credibly threaten escalation against B, B may be able to counterthreaten escalation against A, thereby forcing A to back down. A key factor in successful coercion is therefore "escalation dominance": the ability to increase the costs while denying the adversary opportunity to neutralize those costs or counterescalate. Escalation does not involve minor changes in tactics but rather a shift—recognized as such by the parties concerned—in the scale and scope of a conflict.[1] For example, the U.S. attack on

---

[1]Smoke (1977), pp. 32–33. Smoke draws on Thomas Schelling's idea of a saliency—objective and identifiable points whose crossing is recognized as significant by the parties involved. Herman Kahn introduced the important idea of recognizing that escalation must be understood in context. As Kahn notes, in an escalation situation, "either side can win if it increases its efforts in some way, *provided that the other side*

Bosnian Serb forces in 1995 demonstrated escalation dominance—the attack went far beyond previous strikes and left the adversary unable to make a corresponding increase in the scope and scale of its effort.

The realistic possibility of nuclear use poses the most vivid example of how escalation can lead to successful coercion. During the Cuban Missile Crisis, the Korean War, and even Desert Storm, the possibility of nuclear use—by one side—contributed to successful coercive diplomacy. As the confrontation over Cuba unfolded, U.S. intelligence informed the Kennedy administration that Soviet nuclear forces were in a poor state of preparedness and that the United States could, if necessary, launch a devastating first strike with a low probability of a robust Soviet response. This dominance allowed Kennedy to stake out a demanding public profile; he knew that the costs of escalation would weigh more heavily on Moscow.[2]   In the Korean War, the North agreed to accept talks leading to the continued partition of the country in part because of the election of President Eisenhower, who threatened the use of nuclear weapons to end the conflict.[3]   This threat, while probably not sufficient by itself to bring about peace, contributed to the North Korean and Chinese decision to seek peace. Similarly, the threat of nuclear use did not lead Iraq to capitulate to the United States after the invasion of Kuwait, but it did play a role as a deterrent in leading Saddam to refrain from chemical weapons use during the campaign.[4]   In each of these cases, the adversary lacked the option for matching U.S. escalation to nuclear weapons.

Escalation dominance can be achieved by conventional military might as well. At times, one side may have such a preponderance of force that the adversary perceives little choice but to concede. In 1961, the Trujillo family leaders in the Dominican Republic recognized that the U.S. show of force might be a prelude to an actual invasion—a fear reinforced by repeated U.S. interventions in the re-

---

did not negate the increase by increasing its own efforts." Kahn (1965), p. 3 (emphasis in original).

[2]Fursenko and Naftali (1998); George and Simons (1994), p. 125.

[3]Clodfelter (1989).

[4]Baker (1995), p. 359, and Iraqi News Agency Broadcasts (1991).

gion and by the deployment of a U.S. task force to the area.[5] More recently, in Bosnia, Operation Deliberate Force succeeded in part because Serb leaders understood that the U.S. air strikes could increase in number and scope and inflict even greater damage on their forces.

Rather than threatening to employ higher levels of violence, coercers often threaten to maintain a steady level, utilizing the prospect of mounting aggregate costs to influence adversary decisionmaking. Although such a strategy would be effective against a perfectly rational adversary, states and leaders facing these types of threats are often unduly optimistic about the consequences of their actions and will attempt to wait out coercion when possible.[6] The result is a long-delayed reaction, though sometimes ultimately a favorable one. Israeli operations against Palestinian terrorists in Jordan and Iranian support of Kurdish insurgents in Iraq took years to bear fruit. Such operations did not bring about a rapid change in the adversary's decisionmaking and, had they been halted, would have gone down in history as failures of coercion. The Jordanian and Iraqi governments slowly realized that they could not prevent the intervention and that the costs imposed might lead to their loss of power. On a lesser scale, the U.S. Linebacker II operations in Vietnam succeeded in part because the North Vietnamese perceived them as sustainable. That is, Hanoi recognized that Washington could continue its massive air campaign in the face of a paucity of North Vietnamese air defense assets.

The credible ability to maintain or raise the level of military force allows a coercer to manipulate the adversary's expected costs. Adversaries are not passive, however, and they regularly try to turn the tables on the coercing power. In so doing, they in effect coerce the coercer, imposing costs and threatening to impose more, until

---

[5]Slater (1978).

[6]For more on such biases in decisionmaking, see Plous (1993), pp. 15–18; Janis (1982); Camerer (1995), and Kahneman and Tversky (1979). Jack Levy notes the importance of the "endowment effect" on political decisionmaking—a phenomenon that makes people overvalue their current possessions (and thus incur more costs or gain fewer benefits than would a perfectly rational individual) (Levy, 1997, p. 89). Levy concludes that coercion becomes harder as a result (p. 93). Paul R. Pillar notes that individuals often are willing to suffer more pain as a conflict nears an end, buoyed by the belief that suffering will soon end (Pillar, 1990, p. 253).

the initial coercer backs down from its demands. Perhaps the best
example of this is the Egyptian response to various coercive threats
faced during the Nasser era. When Israel, France, and Britain attacked in 1956 following Nasser's nationalization of the Suez Canal,
Egypt scuttled ships in the canal, closing it to traffic. Nasser thereby
sought to stop the invading powers by fulfilling their worst fears: the
closing of the canal. During the War of Attrition, Israeli deep-
bombing strikes proved highly effective, devastating Egyptian air
defense and artillery positions and creating consternation among
Egyptian leaders. Rather than concede, however, the Egyptians
turned to the Soviet Union for assistance. Moscow not only sent
more air defense assets, it also provided Soviet crews to man them
and Soviet personnel to fly counterair operations. Israel thus faced a
superpower's wrath when it continued its coercive campaign, leading it eventually to accept a cease-fire that in essence hindered its
command of the air over the canal zone.[7]

Adversaries often try to impose costs on the coercer by causing casualties that, by themselves, do not impede military effectiveness but
cause political turmoil in the coercing power. The Chechens, for example, struck deep into Russian territory, seizing hospitals and otherwise sowing mayhem on nonmilitary targets in order to mobilize
Russian sentiment against the war. Similarly, in Somalia General
Aideed, who drew support from the Habar Gidir subclan, deliberately
tried to create U.S. and UN casualties in order to force the United
States out of Somalia. In short, coercers seeking escalation
dominance must think in terms of simultaneous and conflicting coercive strategies, where the adversary attempts to coerce the coercing power in turn.

In addition to issuing counterthreats, an adversary facing a coercive
threat will often take a variety of steps to neutralize a coercer's ability
to escalate. Adversaries use a number of countermeasures. During
the Korean War, the North Koreans lowered water levels to prevent
U.S. destruction of irrigation dikes from causing widespread flood
damage. In addition, the Koreans employed quick construction

---

[7]Egypt, with Soviet assistance, then successfully set up a SAM box near the Suez Canal,
effectively denying Israel the control of the air it had enjoyed in the past. As Jonathan
Shimshoni notes, "And so, admit it or not, and despite the appearance of a draw, Israel
had lost her first war." Shimshoni (1998), p. 170.

brigades that specialized in rapidly repairing damaged facilities. In general, governments have proven skilled at diverting resources from civilian projects and from less critical military activities, making it harder to escalate pressure through stopping enemy military production.[8]

Different regimes have different options when confronted with a coercive threat. Responses are shaped and limited by regime institutions, ideology, and base of support. Some may seek to rally the nation, bolstering their status by defying the coercive power. Others may increase repression, fearing that the coercion campaign might increase popular dissent. Not all adversaries are equally skilled, but deception, operational substitutes, and other steps to resist the damage done by coercion are almost always attempted. Even Saddam's Iraq, an adversary not known for its brilliant deception and subtlety, has used decoys, created backups for communications links, and removed equipment before strikes occurred. In response to U.S. encouragement of internal dissent in Iraq in the years since Operation Desert Storm, Saddam brutally cracked down on Kurdish and Shi'a areas, killing thousands and causing massive refugee flows, thereby reducing the threat of unrest. These countermeasures offset conditions favorable to coercion and exploited conditions unfavorable to coercion.

In addition to operational steps designed to neutralize the threat of escalation, adversaries employ political tactics to offset the coercer's military dominance. A common technique is the exploitation of civilian suffering. Although U.S. sensitivity to an adversary's casualties is often overstated, U.S. decisionmakers appear increasingly concerned that the deaths of enemy civilians will lead to a collapse in a coercion effort (see Chapter Four). Toward the end of Operation Desert Storm, U.S. political leaders placed restrictions on bombings of targets in civilian areas because of deaths at the Al Firdos bunker after a U.S. air strike. Saddam dramatized these deaths in the international media, hoping to create a backlash in the United States and among its allies. Although this effort failed to disrupt the entire campaign or even to generate sympathy among the U.S. people, it did

---

[8]Pape (1996). A more comprehensive overview of how governments respond to shortages during war can be found in Olson (1963).

lead U.S. commanders to curtail the air strikes on Baghdad.[9] Saddam has subsequently succeeded in having sanctions reduced for "humanitarian reasons," as the United States seeks to avoid any perception, especially among its Arab partners, that it is targeting innocent Iraqi civilians along with the Ba'ath regime. The exploitation of suffering imposes diplomatic and domestic political costs on the coercing power. It also can increase support for the adversary among key constituencies, which rally against the killings.

U.S. air power can play a major role in reducing an adversary's ability to escalate, and thereby help to secure escalation dominance. Because of the growing ability of air power to attrit and destroy an adversary's fielded forces, adversaries have fewer options for using their conventional forces to strike out. Saddam Hussein, for example, refrained from using his conventional forces in 1994 against Kuwait—despite reports that he was considering military action to compel the UN to lift sanctions—because the prompt U.S. response (Operation Vigilant Warrior) produced sufficient fear that the United States would destroy his army.

The potency, mobility, and flexibility of air power give planners a range of escalatory options when designing coercive strategies, without placing friendly forces substantially at risk. At the same time, precision strikes can reduce enemy casualties, and therefore reduce fears of U.S. or international popular revulsion which in the past have proven a key constraint on escalation. These advantages were evident in Operation Deliberate Force, where NATO air power could strike accurately at Serb military assets with relative impunity.[10]

Escalation dominance is a product of the three elements outlined above: the capacity and will for higher levels of coercive force, the ability to prevent an adversary from escalating, and the ability to neutralize an adversary's counter-coercive measures. When escalation dominance is secured, a coercer often has substantial control over the level of costs the adversary expects; the coercer can turn up the dial as necessary. But this is just one way in which a coercer can manipulate an adversary's cost-benefit calculations.

---

[9]Arkin (1997); Gordon and Trainor (1994), p. 326.

[10]Operation Deliberate Force, as discussed later in this report, also illustrates some limits of air power as a coercive instrument.

Threatening to Defeat an Adversary's Strategy

Coercion is more effective when it renders impotent an adversary's strategy for winning or, in coercive terms, gaining the desired benefits. As Robert Pape argues, to be effective "the coercer must exploit the particular vulnerabilities of the opponent's specific strategy."[11] Such a "denial" strategy focuses on the benefits side of the coercion equation, reducing the incentives for an adversary to engage in the hostile behavior.[12] A denial strategy at times blurs with "brute force"; both usually seek to defeat an adversary's military, but "denial" focuses on convincing an adversary that future benefits will not be gained through military means, whereas more conventional warfighting focuses on physically stopping an adversary regardless of whether its leadership believes it can fight on.[13]

When its strategy for victory is thwarted, an adversary is more likely to come to the negotiating table. Argentina recognized that it could not hold the Falkland Islands after British military successes. As a result, it ended its resistance even though it could have continued the struggle for years, because Britain did not have the will and capacity to invade and occupy Argentina proper. Attacks on Chinese logistics during the Korean conflict proved devastating (a greater percentage of truck drivers died than did frontline troops), helping stop Chinese thrusts and reducing China's ability to gain victory. Once the adversary's leadership realized that victory was impossible—a process that took years in China's case—it proved more willing to make conces-

---

[11]Pape (1996), p. 30. We think denial is one effective strategy, but not the only one.

[12]According to Pape, "Using air power for *denial* entails smashing enemy military forces, weakening them to the point where friendly ground forces can seize disputed territories without suffering unacceptable losses. Denial strategies seek to thwart the enemy's military strategy for taking or holding its territorial objectives, compelling concessions to avoid futile expenditures of further resources." Pape (1997/1998b), p. 97. (Emphasis in original.)

[13]Denial in coercion is not the same as denial in war. Coercive denial hinges on the perception that benefits will not be achieved; denial by warfighting rests on making such a possibility a reality. In Operation Deliberate Force, the United States made it clear that any benefits the Bosnian Serbs hoped to achieve would not be realized because the United States would resist further offensive operations; in fact, the Serbs risked losing more territory to the Croats and Muslims if they continued fighting. In Operation Desert Storm, in contrast, the United States simply pushed Iraqi forces out of Kuwait, leaving them with no choice but to comply.

sions.[14] Similarly, the Bosnian Serb leadership recognized that
NATO air attacks would make it difficult for the Serbs to defeat the
ongoing Croatian land offensive in 1995. Iraq in 1975 also recognized, after over a year of unsuccessful fighting, that it could not defeat Kurdish insurgents as long as they had Iran's backing. Thus,
Baghdad agreed to Iran's demands about their contested border.

The key to successful denial is to defeat the enemy's strategy for victory, not simply its conventional military operations. To force an adversary to recognize a military stalemate or defeat, denial campaigns
generally attack military production, interdict supplies to the battlefield, shatter enemy air defenses, disrupt communication and command, and defeat fielded forces.[15] The degree to which these can
effectively alter behavior depends on the nature of the adversary and
its strategy. Pape argues that the Rolling Thunder campaign, as well
as the U.S. interdiction efforts in Laos and during the Korean War,
failed in large part because the resource needs of the adversary's
fighters were limited. Although the United States devastated the
transportation grid and hindered throughput, the guerrillas and
soldiers required relatively few supplies, allowing them to use the
degraded transportation network.[16] On the other hand, the Linebacker campaigns in Vietnam succeeded because the North Vietnamese had switched to a conventional military strategy. U.S. air
power proved highly effective at cutting off the supplies and infrastructure necessary for conventional operations. After failing to
sustain conventional operations in the South, Hanoi realized that
military success depended on removing the United States, and
particularly the U.S. Air Force, from the conflict.[17]

Successful denial is impossible when an adversary's strategy cannot
be thwarted by military force. Saddam Hussein's 1997–1998 chal-

---

[14]Yu (1998), p. 9 (internet version).

[15]Pape (1997/1998b), p. 97.

[16]Mark Clodfelter argues that air power was ineffective when North Vietnam
employed a guerrilla strategy, but was effective when North Vietnam used
conventional military operations. "Because of revamped American political objectives
and the North's decision to wage conventional war, Linebacker proved more effective
than Rolling Thunder in furthering U.S. goals in Vietnam." Clodfelter (1989), p. 148.
See also Pape (1996), pp. 193–194.

[17]Pape (1996).

lenge of the UNSCOM inspection regime, for example, was not susceptible to military denial because Iraq's objectives—an end to sanctions, greater prestige, and splitting the anti-Iraq coalition—could not be denied through military force.[18] Indeed, the use of air strikes or other military measures might have even aided Iraq's political strategy. Some military strategies are difficult to counter using limited force, particularly air power.

Operations that threaten to deny an adversary its desired gains, even if of only limited value on their own, can combine with escalation dominance or other positive factors to make favorable outcomes more likely. For example, attacking the Iraqi Republican Guard not only reduced this force's ability to threaten Iraq's neighbors, thus preventing Saddam from invading Kuwait, but also threatened Saddam's strategy for maintaining control over the military and restraining insurgencies, because the Guard played a vital counterinsurgency and countercoup role. Thus, attacks on the Guard could compel concessions even though attempts at denial seemed unlikely to work in isolation.[19]

Coercion in Context: Magnifying Third-Party Threats

As the above examples highlight, coercive threats often rely on the direct effect of military strikes, including the destruction of conventional forces. But coercive strikes can also have indirect effects that shape an adversary's cost-benefit calculus. The cases studied indicate that coercion is more likely to succeed when coercive threats magnify other threats to the adversary, such as external military and internal threats.

Successful coercive operations can magnify an external threat by reducing the ability of the adversary to defend against the third party. In such cases, the adversary fears not only the immediate punishment imposed but also further losses at the hands of a third party.

---

[18]Arguably, Operation Desert Fox (December 1998) was a brute force effort to degrade Iraqi weapons of mass destruction (WMD) capabilities after coercive threats failed to gain full Iraqi compliance with UNSCOM. Desert Fox had a coercive component to it as well, however, because it sought to encourage a coup or rebellion and to bolster U.S. credibility in the region.

[19]See Byman, Pollack, and Waxman (1998).

The external threat thus acts as a force multiplier, dramatically increasing the effect on cost-benefit calculations.

Operation Deliberate Force illustrates how a third-party threat can magnify the effect of a coercive campaign. For several years, the Bosnian Serbs had ignored UN and NATO ultimata. NATO's September 1995 air strikes on Bosnian Serb forces not only hurt the Bosnian Serbs directly, but they posed the risk that Bosnian Muslim and Croat forces would make further advances at the Serbs' expense. Recent Muslim and Croat battlefield successes, particularly the successful Croat offensives against the Serbs in western Slavonia and in the Krajina, forced the Serbs to consider that defiance of the United Nations might lead to defeat at the hands of their enemies, not just further air strikes. As one post-Deliberate Force operations analysis concluded: "Hitting communication nodes, weapons and ammunition storage areas, and lines of communication took away Serb mobility and did not allow them to respond to . . . offensives elsewhere in Bosnia."[20] U.S. strikes altered the local military balance and exposed vulnerabilities in Serb defensive capabilities.

External foes are not the only type of third-party threat that coercive strategies can exploit. Coercive diplomacy may succeed by fostering internal instability in the adversary's country. In such cases, the costs imposed are not directly related to the coercive campaign but rather stem from the adversary's society itself. As with other coercive tools, magnifying an internal threat raises the costs of resistance, but it does so by focusing on the weaknesses of the regime.

Internal security is of overriding concern to developing states.[21] The basis of authority of a particular regime, which is the core of internal security, will therefore shape the perceived costs and benefits of alternative courses of action. A regime that depends on the army to stay in power is likely to be more vulnerable to attacks on its armed forces than a government that has broader popular support. Similarly, a culture that places a high value on honor and face might be more susceptible to attacks that had little operational impact but demonstrated the regime's weakness.

---

[20]Beale (1997), p. 37.

[21]See Ayoob (1991) and David (1991).

Israel's cross-border operations against Jordan in the 1950s and again in the late 1960s highlight the potency of magnifying internal threats when such a strategy is crafted to fit a particular adversary regime. Plagued by Palestinian cross-border attacks from refugee camps in Jordan, Israel engaged in regular strikes inside Jordan. This retaliation led Palestinian groups to stay well armed and damaged the credibility of the Jordanian government. Thus, Israel's operations raised the specter of unrest in Jordan, as local guerrillas became more active, better armed, and more critical of the Hashemite regime. King Hussein, while outwardly professing defiance of Israel, instructed his army to crack down on cross-border operations. When the Palestinian operations began anew after the 1967 war, Hussein ordered his army to suppress all Palestinian guerrillas, leading to a bloody battle in 1970 that forced the Palestinian guerrillas to flee to Lebanon. Because Hussein feared internal unrest and sought to integrate Jordan, Israeli operations threatened to impose unacceptable costs.[22] These costs led—after repeated strikes—to the end of Palestinian attacks from Jordan.

To a lesser extent, the Iran-Iraq border dispute also illustrates the effectiveness of a coercive strategy that threatens internal unrest. Iranian support for Iraqi Kurds threatened one of the Ba'ath regime's chief objectives—integrating Iraq under an Arab nationalist regime. The Kurdish dispute in the past had led to the fall of several Iraqi governments, and the Ba'ath regime recognized that an imperfect peace was better than continued fighting. The Ba'ath leadership therefore acceded to Iranian demands over the disputed border.[23]

Public reaction to coercive threats is extremely unpredictable, however, and the finding that coercion often succeeds when it magnifies an internal threat does not mean that attacks should be conducted so as to undermine civilian morale. Indeed, a recurring historical lesson, particularly since the advent of air power, has been that attempts to force an adversary's hand by targeting its populace's will to resist may well backfire.

---

[22]See Shimshoni (1988), pp. 37–51, and Morris (1997), pp. 100–101, for information on 1950s operations. See Dupuy (1992), pp. 378–381, for information on the Palestinian guerrillas and the crisis in Jordan in 1968–1970.

[23]Marr (1985); Chubin and Tripp (1988), p. 23.

A common response to coercion is that an adversary's determination stiffens as both the leadership and the country as a whole unite against the coercer. Indeed, the coercive campaign itself tends to raise the cost of compliance for an adversary's leadership if it provokes a hostile backlash. Russian attempts to bomb the Chechens and the Afghans into submission simply led to unified defiance— even residents who formerly favored peaceful solutions, or favored fighting each other, united to expel the invader. U.S. operations in Somalia, although humanitarian in nature, faced a similar problem when the United States attacked and killed several leaders of Aideed's clan. Although the clan leaders had been highly critical of Aideed's confrontational stance toward the United States, they united when faced with a direct outside threat. Such defiance, moreover, can enhance a leader's stature even when the leader cannot produce military success. Egyptian President Nasser lost the Suez War, but he became more popular than ever with his unbowing stance toward Israel, France, and Britain. As Donald Neff notes about the attacks on Egypt:

> The bombings, though carefully kept away from civilian targets, were nonetheless having the same counterproductive result that they had had in London during the Nazi aerial war. They were stiffening civilian resolve and morale. During the rest of the crisis, Nasser was greeted by shouts repeating his defiant motto as he drove through Cairo streets.[24]

When coercive operations threaten to foster instability, whether wittingly or unwittingly, target regimes may be prepared to respond. Regimes faced with threats of a coup frequently do not hesitate to purge officer corps. Indeed, they may do so regularly. Since the end of Operation Desert Storm, Iraq has repeatedly arrested and killed senior army leaders suspected of plotting against the leadership. If widespread domestic unrest appears likely, regimes will increase the police presence, order mass arrests, and even slaughter potential opposition members to preserve their power. In Iran during the Iran-Iraq war, the clerical regime regularly arrested, tortured, and killed suspected opposition members. As the United States is pitted against authoritarian regimes, it will find itself confronted by

---

[24]Neff (1981), p. 393.

governments skilled at maintaining order. Although these regimes at times have little domestic support, their police and intelligence services can prevent instability from toppling the leadership. The potential costs that coercive strategies can inflict are thus reduced.

Striking the proper balance between provoking unrest and inflaming nationalism is difficult, and the historical record does not offer a clear set of predictive conditions. The cases examined do, however, point to the need to anticipate obvious counter-coercive strategies designed to strengthen an adversary regime vis-à-vis its opponents. Even though an adversary's leadership may seek to bolster its standing by resisting threats, carefully constructed coercive strategies can magnify a variety of third-party threats and success-fully compel adversary compliance without the need for overwhelming military force by the coercer.

COMMON CHALLENGES IN COERCIVE OPERATIONS

The cases examined revealed several sets of challenges that routinely plague coercive operations: intelligence and estimation challenges, credibility challenges, and feasibility challenges. Even when a co-ercer expects to dominate the escalation contest, defeat an adver-sary's military strategy, or magnify another threat, any one of these challenges can wreck a coercive strategy.

Intelligence and Estimation Challenges

Coercion is, of course, more likely to succeed when intelligence is plentiful and accurate. Indeed, any military or foreign policy strategy depends on timely and precise pictures of the situation and the ad-versary's order of battle. Coercion is about altering costs and bene-fits, so knowledge of relative strengths and weaknesses is an obvious prerequisite for selecting inputs on which to focus.[25] But coercion is

---

[25]Intelligence coups played an important role in several successful instances of coercion. In the Cuban Missile Crisis, the intelligence community's early discovery of the missile facilities on Cuba and its knowledge that the Soviet missiles were in a poor state of readiness contributed to successful coercion. In a smaller example, the U.S. reflagging operations in the Gulf in 1987 and 1988 were successful in part because the United States uncovered, and recorded, Iranian minelaying operations, thus making it

a psychological phenomenon as well. It works by manipulating perceptions to affect human and state decisionmaking. Although rapidly evolving information technologies may offer improvements in the U.S. ability to locate and target valuable assets or fragilities in an adversary's defense, the cases studied illuminate the need for sophisticated understanding of the adversary regime's objectives, decisionmaking apparatus, and the ways in which it will react to certain forms of threat—knowledge that is often best gleaned from agents, diplomats, or journalists, not technical means. Even when governments share similar objectives, they often order them differently. A regime based on a pan-Islamic ideology, for example, may place a higher value on supporting co-religionists abroad than would a regime supporting insurgents for purely instrumental reasons. Similarly, a regime may not value all territory equally, instead caring about a particular region more for geographic, strategic, or other reasons. In sum, successful coercion requires not only the collection of accurate data but a sophisticated integration of those data to generate the total profile of the adversary necessary to design appropriate coercive threats.

The primary intelligence challenge is to discern the interests and nature of the adversary. When the benefits to the adversary of resistance are relatively low, less is required to compel it to abandon its actions. Israeli operations against Jordan, for example, capitalized on the Hashemite regime's at-best lukewarm support for the Palestinian cause, particularly when compared with the stability of the monarchy itself. Similarly, the Linebacker II operations in Vietnam succeeded because the United States demanded relatively little of Hanoi. The United States allowed the North Vietnamese to continue their military presence in the region. Moreover, the Communist regime saw any concession as a temporary step, an instrumental way to force the United States out of Vietnam politically. Similarly, although the Bosnian Serbs did not seek negotiations, the Dayton Accords allowed them considerable control over much of the Bosnian territory that they had retained in the ground war, thus asking relatively little of them. By contrast, when an adversary is fighting for the defense of its homeland or other vital goals, coercion becomes

---

impossible for Tehran to deny responsibility. This knowledge allowed the United States to retaliate militarily while retaining the support of the Gulf states.

far more difficult and at times impossible. In such cases, the costs of concessions may be too high, making continued defiance the better option for the adversary.

The value of air power and other military instruments is directly related to the quality of the intelligence available. For denial strategies to succeed, the United States must have a comprehensive knowledge of the adversary's fielded forces. This intelligence must provide the coercer with enough information about the adversary's goals to design a strategy tailored to thwart them. During the Vietnam War, poor U.S. understanding of the relationship between destroying much of the North's transportation network and the needs of the guerrillas in the South led the United States to focus considerable energy on degrading throughput in its attempt to coerce Hanoi, when in fact the enemy forces in the South needed only small quantities of materiel. Sophisticated intelligence is especially necessary to exploit regime instability. To know which opposition group to aid, or which segment of the regime's forces are devoted to suppressing unrest, the coercing power must have keen area expertise and individuals in contact with opposition groups.

A common intelligence deficiency—often translating to a failure of coercion—is poor assessment of an adversary's determination, creativity, and resilience. Coercers frequently see their adversaries as poorly motivated, waiting to collapse after a brief military campaign. Iran and Iraq both believed that the other was on the verge of collapse and that attacks on cities would cause their rival's government to collapse. Experience, however, shows bitterly that adversaries often are far more motivated than the coercing power. In Somalia, for example, 18 U.S. servicemen died in the Mogadishu firefight, while perhaps 1000 Somalis died. Yet it was the United States that yielded. The Russians similarly attempted to overwhelm the Chechens by bombing and shelling populated areas, believing they could intimidate the Chechens into peace. Russian military leaders dismissed the fighting capacity of the Chechen resistance, leading to a debacle.[26] If anything, the Russian attacks only increased the adversary's determination.

---

[26]Lambeth (1996), p. 278.

Casualty sensitivity and other elements of "cost" usually vary from country to country. This is not to repeat the canard that some cultures value human life less than others. Rather, some governments are more willing to sacrifice their people and are less receptive to popular complaints about casualties. Nazi Germany and imperial Japan, for example, accepted the deaths of 330,000 and 900,000 citizens, respectively, from the Allied bombing campaigns. (Germany and Japan today are highly sensitive to casualties.)[27]

A simple example of how regime attitudes toward the population can affect the success or failure of coercion can be found in the repeated U.S. attempts to coerce Iraq. Since 1990, Iraq has suffered under perhaps the most restrictive economic sanctions ever imposed on a major power. Thousands of Iraqis have died, and millions have seen their standard of living plummet. The well-being of the Iraqi people, however, is not a priority of the Ba'ath regime. Thus, the undeniable damage inflicted on the Iraqi people has not threatened key interests or regime priorities. Indeed, Saddam recognized that capitulating to the United States could cause him greater political problems than being defeated by the United States. Drawing from Nasser's experience in 1956, Saddam understood that defeat by a superpower can magnify a leader's stature. Capitulation, however, might hurt his claim to the mantle of Arab leadership and lead to increased internal dissent among his core constituency. Thus, the expected costs of complying with coercion have outweighed the expected costs associated with the coercive threats.

In contrast, in 1956 the United States successfully coerced Britain into withdrawing its forces from Egypt, exploiting the close identification of the government with the governed. At the time, the United States had tremendous economic leverage over Britain and threatened not to release funds to uphold the British pound, which otherwise might fall. Faced with the possibility of an economic crisis (though small by comparison to what Iraq has suffered), Prime Minister Eden withdrew British troops. Moreover, although the Suez crisis cost Eden politically, by itself it was not as high a regime priority as maintaining the value of the pound. The costs of mild U.S.

---

[27]Pape (1996), pp. 128–129, 254. The figure for Japan includes the casualties caused by the two atomic bombs dropped on Japan.

coercion were high, whereas the benefits of continuing on with the Suez expedition were small.[28]

A problem related to underestimating the adversary's resilience is overestimating the coercer's effectiveness or the value of the target. In so doing, the coercer assumes that the costs being imposed are higher than they actually are. Practitioners often ignore the robustness of target sets, assuming that the enemy will not be able to repair destroyed facilities or overstating the damage done to a target. During World War II, for example, Army Air Force planners sought to cripple Nazi Germany by destroying its industrial capacity. The intention was to destroy critical nodes, such as aircraft engine and ball-bearing plants, to prevent completion of aircraft and other final products. Planners, however, grossly underestimated the robustness of German war production facilities and infrastructure. Factories weathered repeated strikes. Even when air strikes destroyed part of a factory, much of the capacity (both capital and labor) survived to continue use.[29] Overestimating success similarly occurs in judging an adversary's political resilience. In the Chechen conflict, Moscow thought that the occupation of Grozny would lead to the rebels' collapse. Instead, it led them to form guerrilla bands in the countryside.

Often the intelligence necessary to destroy the adversary's critical assets is lacking. The "critical node" approach used in World War II relied on understanding the entire German industrial system and then finding the vulnerable links. Such a systemwide approach, while intellectually appealing, is difficult to implement because even limited intelligence gaps—the rule, not the exception in war—can cause the campaign as a whole to fail. These intelligence gaps sometimes occur because of "mirror imaging" the coercer's vulnerabilities onto its enemy. In Operation Desert Storm, U.S. planners overstated the importance of destroying Iraq's real-time command and control communications links.[30] The U.S. military is heavily dependent on modern communications to conduct operations. Third World militaries, however, may rely on far more primitive systems. Saddam, for

---

[28]Neff (1981), pp. 409–410; Kirshner (1995), pp. 63–82.

[29]Olson (1962); Mierzejewski (1988).

[30]The USAF devoted 580 strikes against command, control, and communications targets during the campaign. Keaney and Cohen (1993), pp. 66–71.

example, regularly used couriers to communicate with his troops and, in any event, the regime appears to have remained in touch with the Kuwait theater.[31] Quite frequently, Third World military officials do not coordinate their activities carefully, a tendency that reduces their operational effectiveness but leaves them less vulnerable to threats to their communications network.

Contributing to these intelligence challenges is a tendency for planners to define success as destroying a target rather than as inducing the desired behavior. During Operation Desert Storm, U.S. air strikes effectively destroyed Iraqi electric power generation. The regime, however, was willing to pay that price; the effect on Iraq's ability to wage war, or on the people's desire to rebel, in retrospect appears negligible.[32] Similarly, in World War II the United States destroyed Japanese factories that were already idle because of lack of fuel and raw materials resulting from the U.S. naval blockade. Japanese production therefore did not change.[33] These operations were successful from a military standpoint, but they failed to bring about the desired reaction from the adversary. Campaigns often take on a life of their own and become divorced from the political and behavioral objectives sought.

Misperceptions and Coercion

A primary component of intelligence is recognizing the importance of perceptions as well as objective reality. Adversary perceptions are shaped by a variety of factors, including leadership dynamics, regime type, and culture, all of which can affect the way the adversary views and calculates the costs and benefits of resistance.[34] The failure to

---

[31]Arkin (1997), pp. 4–6.

[32]Keaney and Cohen (1993), pp. 74–75.

[33]Many unnecessary or unproductive strikes are an inherent part of the friction of war. As noted above, damage to facilities may be unknown, and the effects of particular strikes unclear. In such circumstances, redundant attacks are preferable, if resources are available.

[34]There is a vast literature assessing how decisionmaking procedures and cognitive biases skew perceptions, particularly during crises. Useful reviews can be found in Janis (1982); Camerer (1995); *Psychological Dimensions of War* (1990); Booth (1979); and Levy (1997). For an asessment of how air operations affect opponents psychologically, see Hosmer (1996).

understand perceptions can erode the effectiveness of coercive strategies.

Although perceptual distortions are common, certain attributes make them endemic to particular regime types. Isolated leaders, for example, are more likely to misperceive a situation than are leaders with a coterie of well-informed, confident advisers. Different cultures may place different interpretations on surrender, negotiation, or other responses.

A number of studies have shown that misunderstandings of cues and perceptual distortions exacerbated by crisis situations have contributed to breakdowns in deterrence.[35] Recent revelations from the Cuban Missile Crisis similarly show that opposing leaderships may grossly misconstrue signals, potentially contributing to inadvertent escalation. For example, U.S. leaders saw the missile deployment as the first part of a campaign to weaken the U.S. position in Berlin; the Soviets, in reality, were primarily trying to ensure Cuban independence.[36]

Similarly, it is often difficult to send a clear and consistent message through the threat and use of force. Rolling Thunder planners, for example, intended to smoothly escalate the bombing to demonstrate to Hanoi the increasing costs of resistance, but weather and operational friction resulted instead in almost spasmodic escalation. Today, a stray precision munition could send a message regarding U.S. intent that was quite different from what U.S. decisionmakers sought.[37]

Credibility Challenges

A problem common to failures of coercion—one identified in many studies on the subject—is doubts about a coercer's credibility. Successful coercion depends on the threat of future costs. When an

---

[35]Good overviews of various misperceptions common to decisionmakers can be found in Jervis (1976), Khong (1992), and Janis (1982).

[36]Fursenko and Naftali (1998).

[37]Thies (1980) reviews how aerial coercion can send the wrong message. We thank Karl Mueller for his suggestion concerning the impact of a stray precision-guided munition.

adversary doubts the coercer can escalate—or even sustain—operations, the perceived costs of defiance fall. The North Vietnamese, having succeeded in their long war with the French, rightly believed that the United States would eventually weary of the conflict and withdraw. Similarly, the Chechens recognized that the Russians, who had faced a long and bitter war in Afghanistan, would avoid another imbroglio.[38]

Domestic unrest in the coercer state or disputes among allies can undermine the credibility of threats in the adversary's eyes. The North Vietnamese took heart at domestic demonstrations against the war in the United States, as did the Chechens over the Russian public's opposition to the war in Chechnya.[39] Saddam has become increasingly confident in his resistance to the United Nations as France and Russia appear to be defecting from the once-staunch anti-Iraq operation. Adopting restrictive rules of engagement (ROE) can further undermine the coercer's credibility. All sides in the Bosnian conflict, for example, dismissed initial European peacekeeping forces and even took them hostage at times, in part because these forces operated under severe restrictions on their ability to use force.

Success or failure in conflicts affect the coercer's reputation, and thus its ability to coerce in future crises. Algerian insurgents planned a long war for liberation against France, believing that the Vietnam experience proved that France could not tolerate a long guerrilla war. Saddam Hussein rejected calls to pull out of Kuwait because he believed that the United States could not stomach the casualties of a ground war. The very real danger of U.S. armed strikes was discounted by Saddam. He was particularly scornful of air power. Shortly after invading Kuwait, he declared, "The United States relies on the Air Force and the Air Force has never been a decisive factor in

---

[38]Immediately before the invasion, a number of Russian generals, including Deputy Defense Minister Boris Gromov, publicly questioned the ability of Russian military forces to quell the revolt. Such predictions echoed in Chechen warnings: when Yeltsin mobilized forces against the breakaway republic, Aslan Moskhadov, chief of the Chechen general staff, reportedly announced that, "The North Caucasus will become another Afghanistan for Russia." (Kohan, 1994).

[39]Clodfelter (1989).

the history of wars."[40] Saddam was mistaken in his assessments, but his perceptual errors diminished the coercive potential of U.S. military prowess.

Doubts may arise not only about a coercer's ability to sustain operations but about its willingness to end them. Hezbollah, for example, may have resisted Israeli pressures in part because Hezbollah leaders believed that Israel would remain hostile to the Islamic movement regardless of any concessions made. They therefore believed that halting guerrilla operations against Israeli targets would not lead Israel to end strikes against the movement. If a coercer intends to combine both threats and positive inducements, its ability to manipulate an adversary's cost-benefit calculus will depend on both the coercer's credibility to carry out military strikes and also the credibility of its commitments.

Feasibility Challenges

Coercing powers may attempt the impossible, trying to change the decisionmaking calculus of an adversary that cannot, for a variety of reasons, alter its behavior sufficiently to meet the coercing power's demands. In essence, some adversaries cannot be coerced.

An adversary regime may weigh threats to its personal or institutional survival more highly than the loss of territory or other threats from a coercing power. A government may lose power, or at times even the lives of leaders, in response to capitulation—in addition to the actual costs of agreeing to coercion. During World War II, militaristic groups regularly assassinated Japanese politicians who favored compromise. In 1990–1991, Saddam Hussein was willing to give Iran a favorable settlement over their disputed border even though this dispute had cost Iraq hundreds of thousands of lives during its eight-year war with Iran.[41] Saddam had a change of heart because he feared that the U.S.-led coalition jeopardized his very regime, and he hoped that closer relations with Iran would help him

---

[40]Saddam Hussein, August 29, 1990, as quoted in *Essays on Air and Space Power* (1997), p. 141.

[41]In essence, Saddam retreated from his earlier demands that Iraq have full control over the disputed Shatt al-Arab waterway, which forms part of the Iran-Iraq border.

avoid economic and political isolation and provide a haven for military assets that were key to his extended reign.

Thomas Schelling observed in *Arms and Influence* that "Coercion by threat of damage also requires that our interests and our opponent's not be absolutely opposed. . . . Coercion requires finding a bargain, arranging for him to be better off doing what we want—worse off not doing what we want—when he takes the threatened penalty into account."[42] The trouble is, in some cases the perceived costs of giving in are so dreaded that virtually no military threat will compel the adversary to bend. Robert Pape argues that Germany did not surrender to the Allies because German leaders feared occupation by Russia and likely vengeance for atrocities committed by Germany in the East. Thus, the massive bombing campaign against Germany, as well as continued *Wehrmacht* battlefield defeats, could not sway a German leadership that saw continued punishment and likely defeat as preferable to occupation.[43] Prohibitive costs of surrender are particularly likely when ethnic or religious conflicts rage. In such situations, adversaries fear that surrender will entail their subjugation by long-time rivals and may even lead to mass killings. Leaders, in particular, may resist concessions if they came to power by inciting communal passions. For leaders, war is often preferred to peace because it ensures their continued political power.

Fear of concessions is particularly pronounced in nondemocracies, where regimes may lack entrenched, institutional legitimacy and leaders may fear for their lives (and those of their cronies and family) if they are forced from office. In Somalia, Aideed recognized that cooperation with UN forces would marginalize him, eroding a power base that depended heavily on profits gained from looting humanitarian aid. Similarly, Egypt's President Nasser did not want to suffer another loss to Israel during the War of Attrition, leading him to take considerable risks and to continue fighting even after losses mounted. The Argentine *junta* had staked its reputation on wresting the Falklands back from Britain and feared, rightly as it turned out, widespread popular anger if it failed to hold them. More recently, Saddam Hussein has defied the United States and the United

---

[42]Schelling (1966), p. 4.

[43]Pape (1996), p. 310.

Nations in part because he fears that concessions will discredit him at home among key constituencies.[44] More benevolent forms of government, by contrast, are often more open to coercion. British Prime Minister Eden, for example, abandoned the Suez campaign in the face of U.S. pressure even though this discredited his government.[45]

Aside from situations in which political dynamics make the costs of concession prohibitively high, coercive strategies can be doomed if the adversary has limited control over the issue in dispute. The Lebanese government, for example, was not strong enough to force the Palestinian guerrillas operating in Lebanon in the early 1970s to stop their attacks on Israel. Although Israeli attempts at coercion successfully led the Lebanese government to try to crack down on the Palestinians, they failed in the larger, and more important, sense that Palestinian attacks on Israel continued.

The question of control is vexing and requires an assessment of the adversary leadership's stability and strength and the nature of any replacement. Russia discovered to its chagrin that the Chechen rebellion could sustain itself even after Russia assassinated Chechen rebel leader General Dzhokhar Dudayev, a symbol of Chechen defiance. His successors proved equally committed to Chechen independence and hostile to Russia. Russia's attempt to coerce the Chechens failed because the Chechen policy of resistance was less controlled by a single figure than Russian planners supposed.

To avoid the implementation of coercive threats, adversaries often claim to lack control over underlings or proxies. In essence, adversaries seek to avoid the imposition of coercive pressure while avoiding compliance with the coercive demand. During the U.S.-Iran confrontation over Iran's threat to shipping in the Gulf in 1987–1988, Tehran repeatedly claimed that it was not responsible for the mining of the Gulf—a claim refuted when U.S. forces took photos of Iranian craft laying mines. Iran has long professed that the Lebanese Hezbollah, which Iran helped arm, train, and organize, operates be-

---

[44]Byman, Pollack, and Waxman (1998).

[45]Unpopular concessions are less likely to lead to a change of government in a strong totalitarian adversary regime. The lack of popular input into governance and a regime's control of perceptions often prevent a hostile political backlash.

yond the control of Tehran. The leadership of Serbia regularly asserted a lack of control over the Bosnian Serbs. The Serb case was particularly difficult, because the Bosnian Serb political leadership then claimed not to govern certain actions of the Bosnian Serb military or militia units, particularly when seeking to avoid air strikes or increased political pressure.

Pleas of lack of control can be bargaining tactics designed to buy time. At times, however, this lack of control is not feigned: governments often are weak, and local parties may act without seeking permission from above. Exploiting this dislocation of authority, however, can allow adversaries to claim compliance while reaping the benefits of noncompliance.[46]

A final implementation challenge involves incomplete compliance. Adversaries may seek to avoid coercive threats by complying with only part of the coercer's demands, seeking to minimize the costs of compliance. Such partial compliance can create difficulties for the coercer, as both domestic supporters and allies may begin to question the necessity of confrontation after the adversary has conceded to coercive demands. Iraq, for example, has dismantled some of its weapons of mass destruction systems and recognized the Iraq-Kuwait border. Such steps, while far from complete compliance with all relevant UN resolutions as demanded by the United States, have led key members of the anti-Iraq coalition such as Russia and France to question the wisdom of continued sanctions. Similarly, Egypt ended attacks on Israeli positions near the Suez Canal in 1970, but violated the cease-fire agreement by moving in air defense assets. Although such movements greatly compromised Israel's security by hindering its control of the air, Israel found it difficult to resume the conflict, both because of fears of casualties at home and reluctance to endure international opprobrium.

Implementation problems can be eased by *providing inducements* for cooperation. The flip-side of threatening an adversary with additional costs arising from escalation or interference with its military strategy is the offering of inducements for compliance. Traditionally, most coercive strategies focus on raising the costs to an adversary of

---

[46]For a more extensive discussion of coercion and the problems of "dislocation of authority," see Waxman (1997b).

continued provocations or on denying benefits of defiance. Inducements reverse this focus. Instead of raising the costs of defiance, inducements increase the value of concessions. When inducements are combined with more traditional forms of coercion, they make success more likely.[47]

The Cuban Missile Crisis represents perhaps the most famous use of inducements as part of coercive diplomacy. In addition to threatening an invasion of Cuba, the Kennedy administration offered to remove U.S. missiles from Turkey and to promise not to invade Cuba— concessions that both allowed Moscow to "save face" and served the Kremlin's broader strategic ambitions.[48] The Linebacker operations in Vietnam succeeded in coercing the South in part because the United States made a major concession to the North, allowing it to retain troops in the South. In Malaya, the British-backed government offered Chinese residents citizenship and land tenure, thus reducing the discrimination that led many to join or support the Communist guerrillas.[49]

Often the value of concessions is ignored or outweighed by competing concerns. Concessions are seen as signs of weakness, and coercers fear that a precedent will be set for bribing states to cooperate. The Saudis became prized victims for radical kidnappers because they were seen as willing to offer up ransoms readily. Concessions, however, can be effective when combined with penalties.[50] Success requires balance. Concessions alone may lead to blackmail, but offering a carrot with the coercive stick often allows adversary regimes to save face with their own constituencies and makes concessions more palatable.

CONCLUSIONS

This chapter has focused on past attempts at coercion, and the lessons derived from these experiences can teach how to coerce more effectively in the future. Although several of the favorable

---

[47]George and Simons (1994); Baldwin (1971), pp. 19–38.

[48]Gaddis (1997).

[49]Mockaitis (1990), pp. 8–9.

[50]Baldwin (1971), pp. 34–36.

conditions identified above are sometimes outside the control of policymakers and military planners, coercive strategies must be designed with them in mind. Political, social, and technological change affect the context in which coercive strategies take place, but much of what is essential to coercion remains constant. Concepts such as escalation dominance, the threat of military denial, and magnification of third-party threats are vital elements of coercive strategies across a range of crises and levels of conflict.

PART 3. COERCIVE DIPLOMACY TODAY

The context for coercive diplomacy today differs from that in the past. Increasingly, the United States is operating as part of multinational coalitions, which decreases U.S. flexibility and can make coercion more difficult. Domestic politics come into play in ways that may differ from past eras and that can further constrain U.S. options. Finally, future challenges may be directed more often against non-state actors—militias, drug rings, and terrorist groups—whose unique characteristics pose additional challenges for coercion. The following three chapters examine these issues in an attempt to understand the context in which the United States will conduct coercive operations today and in the near future.

Chapter Four

DOMESTIC CONSTRAINTS ON COERCION

Part Two examined factors that promote or inhibit successful coercive diplomacy. Occasional references were made in some of the historical vignettes to various domestic influences that complicated coercion or the coercive use of air power. Domestic factors can severely undermine coercive diplomacy by causing decisionmakers to restrict the scope and scale of a military campaign, place limits on escalation, and encourage adversaries to resist U.S. pressure.

This chapter looks at domestic constraints on coercion. It begins by exploring how, and under what conditions, domestic politics can complicate coercive diplomacy by restricting military operations. It then addresses how an adversary might exploit U.S. domestic constraints to counter coercive threats.

DOMESTIC POLITICS AND THE SUCCESS OF COERCIVE DIPLOMACY

The United States may impose constraints on its own military operations that are more binding than those imposed by adversaries, and this can place the United States at a disadvantage in coercive contests. U.S. adversaries are frequently authoritarian regimes that exercise strong control over their legislatures, judiciary, and press, and limit challenges to their authority through a mixture of political suasion, threats of imprisonment, banishment, or even death. In short, many U.S. adversaries have created closed societies that maximize political freedom of action for leaders and minimize political accountability to others. By comparison, political leaders in the United States contend with domestic political pressures that adversary leaders do not face. The constitutional separation or sharing of powers in the U.S. government, and the resulting independence of the Congress and the judiciary, as well as constitutionally protected freedoms for the press, increase the need for U.S. presidents to justify their uses of force. In stark contrast to the latitude and minimal political accountability enjoyed by many adversary leaders, U.S. decisionmakers face domestic constraints that can severely affect their employment of military force.

Common constraints that can have an adverse affect on U.S. coercive efforts include:

- Restrictive objectives and mission statements

- Restrictive rules of engagement

- Public articulation of an exit strategy, including a "date certain" by which time U.S. forces will be withdrawn

- The use of only the least vulnerable—but not necessarily the most appropriate—combat forces in the U.S. inventory for a particular mission (e.g., cruise missiles vs. manned aircraft, stealth vs. nonstealth aircraft, air power vs. ground forces)

- U.S. participation limited to a support rather than combat role

- Imposition of casualty-minimizing or force-protection measures such as restrictions on patrolling or off-duty movements outside of base camps

- U.S. preferences to act in a coalition, both for burden-sharing and to demonstrate that the operation does not flout the will of "the international community"

- Limited availability of concessions that can be offered to an adversary to provide a face-saving outcome, or to provide a more tangible quid pro quo for adversary compliance with the coercive request

As will be described in more detail, these constraints make coercion more difficult and can present adversaries with opportunities for manipulation.

CONSTRAINTS AND THE DEMOCRATIC SYSTEM

U.S. decisionmakers[1] seek support from the Congress, the media, and ultimately the American public for at least three key reasons. First, seeking such support is a key ritual of democratic governance that ensures that matters of war and peace are at least somewhat consultative and consensual in nature. Second, robust domestic support, when given, can strengthen the president's hand in conveying U.S. determination to adversaries abroad, thereby enhancing coercion, or the perception that the United States will prosecute a conflict to its conclusion.[2] Third, robust domestic support will better ensure that other initiatives on the political agenda will not be derailed or held hostage if the military situation turns out to be more costly or prolonged than expected.[3]

Executive officials accordingly make the strongest public case they can to try to achieve robust support from the Congress, media, and public, and they typically seek to mobilize support for their military operations by arguing that

- important national interests and values are at stake, and that the benefits of the military operation are accordingly high;

- the objectives and mission are clearly defined, the prospects for a successful outcome are high, and the situation is unlikely to turn into a "quagmire"; and

- the costs and risks are low, or that they are commensurate with U.S. equities in the situation, and the U.S. military is in any case doing everything it can to deter or minimize casualties.

---

[1]Throughout this chapter, the term "decisionmaker" generally refers to the executive branch, in particular the National Command Authority (the president and secretary of defense).

[2]By *robust* support, we mean support that is relatively insensitive to increasing costs, setbacks on the battlefield, or other factors. This contrasts with *conditional* support, wherein support is contingent on a narrow set of conditions such as low casualties, coalition participation, or other factors.

[3]For example, President Reagan devoted little attention to the issue of Central America in 1981 and 1982 because many on the White House staff felt that it would carry too high a price and cost the president support for more important, primarily domestic, issues. See Kojelis, Reich, Hinckley, and Parry (1993).

The Challenge of Obtaining Domestic Support

Obtaining domestic support for military operations is troublesome, particularly when nonvital interests are threatened. As shown in Table 2, the difficulty for decisionmakers is that even after making their appeals for support, the support they ultimately receive from the Congress, media, and the public for their military operations is frequently highly conditional or, at best, tepid.

The data in Table 2 show that fewer than half initially supported the presence of U.S. troops in Haiti and Bosnia. They also suggest the conditionality of support for Somalia: over the first roughly eight months of the Somalia operation, support collapsed from nearly eight out of ten to about four in ten.[4]

Low or highly conditional support stems from three main sources: beliefs about benefits, prospects for success, and risks and costs, particularly when they are believed to be incommensurate with the stakes or equities.[5]

Table 2

Support for Bosnia, Somalia, and Haiti
(in percentage)

Do you approve or disapprove of the presence of U.S. troops in. . .			
	Approve	Disapprove	No Opinion
Bosnia (12/15–18/95)	41	54	5
Haiti (9/19/94)	46	50	4
Somalia (1/93)	79	17	4
Somalia (9/93)	43	46	11

SOURCE: The Gallup Organization. See Newport (1995).

---

[4]Of particular interest is the fact that the September 1993 data point was *before* the firefight in Mogadishu in October 1993. Put another way, support had collapsed even before this tragic incident.

[5]For an analysis of public opinion on Bosnia that tests this model, see Larson (forthcoming). For a fairly complete listing of public opinion on Bosnia through 1997, see Sobel (1998).

Low Benefits. In many cases, the perceived benefits of the military operation—the importance of the objectives, the national interests, or core values that are engaged—are not considered vital by a majority of congressional leaders or members of the public, and the low perceived benefits result in low support.

Beliefs about the importance of the benefits of an operation are closely related to willingness to support the use of force: in a poll taken on February 7, 1994, those who believed that the United States had a moral responsibility to stop Serb attacks on Sarajevo were greater than two and one-half times more likely (74 percent vs. 28 percent) to support air strikes against Bosnia than those who did not, and those who thought the United States needed to be involved to protect its own interests were almost two times more likely (76 percent vs. 39 percent) to support air strikes against Bosnia than those who did not.[6]

Unfavorable Prospects. Another reason for low support is that many believe that the prospects for success are much lower than the optimistic picture painted by proponents of forceful action. As with beliefs about benefits, support for military action is closely associated with beliefs about its probable efficacy. Those who thought that air strikes would be effective in stopping Serbian attacks on Sarajevo were more than two and one-half times more likely (71 percent vs. 26 percent) to support air strikes in Bosnia than those who did not.[7]

High Costs Relative to the Benefits and Prospects. The third principal reason that support may be low is that the risks or likely costs appear to be too high, given the perceived benefits and prospects for success. Table 3 presents data on the level of support for the Bosnia operation at different hypothesized casualty levels, and shows a decline in support from nearly seven in ten if no American soldiers were killed to 31 percent if 25 were killed.[8] Table 3 also presents data con-

---

[6]Data are drawn from a Gallup poll on July 2, 1994 ($p < .001$ in a chi-square test of independence).

[7]Data are drawn from a Gallup poll on July 2, 1994 ($p < .001$ in a chi-square test of independence).

[8]In many past cases, support for a military operation has declined as a function of the log of the casualties, although the sensitivity to casualties has depended on the perceived benefits and prospects for success. See Larson (1996a, 1996b) for an

firming the relationship between beliefs about support and casualties: it shows that those who believed that the peacekeeping mission was "likely to lead to a long-term commitment involving many casualties" were more than three times (65 percent vs. 19 percent) more likely to oppose the operation than those who thought this was unlikely.

As shown in Table 4, this relationship between support and expected costs frequently manifests itself in a modest preference for air power over more vulnerable ground troops. As the data in the table show, while a razor-thin majority of 50 percent favored the use of air power in Bosnia in February 1994, fully 10 percent fewer were willing to use U.S. and NATO ground troops in the event that air power failed to protect civilians in Sarajevo. (This additional support for air power, while not overwhelming, can mean the difference between the use of force and avoiding intervention when the U.S. interests engaged are not vital.)

Table 3

**Beliefs About Casualties Versus Support for Bosnia, November 27, 1995
(in percentage)**

Now that a peace agreement has been reached by all the groups currently fighting in Bosnia, the Clinton administration plans to contribute U.S. troops to an international peacekeeping force. Do you favor or oppose that?

If the United States sends troops as part of a peacekeeping mission, do you think that it is likely to lead to a long-term commitment in Bosnia involving many casualties, or not?

	Those Favoring Contribution	Those Opposing Contribution
Yes, likely to lead to long-term commitment	35	65
No, not likely	81	19
Don't know/refused	60	40

SOURCE: Gallup poll, 11/27/95.

NOTE: $p < .001$ in a chi-square test of independence.

analysis of the relationship between support and casualties in World War II, Korea, Vietnam, Panama, the Gulf War, and Somalia, among other wars and military operations, and Larson (forthcoming) for an analysis of public opinion data on Bosnia.

Table 4

**Preference for Air Power Over Ground Forces in Bosnia, February 1994
(in percentage)**

As you may know, President Clinton and NATO leaders have threatened air strikes against Serbian military forces that surround Sarajevo if the Serbs do not withdraw their forces from around that city in the next 10 days. The Serbs say they will do this.

—If the Serbs do not do this, would you favor or oppose the U.S. and NATO using planes to bomb the Serbian military positions?

—If air strikes don't make the Serbs withdraw their forces from around Sarajevo, do you favor or oppose sending U.S. and NATO ground troops into Sarajevo to protect the citizens there?

Favor using U.S. and NATO planes to bomb	50
Favor sending U.S. and NATO ground troops	40

SOURCE: Time/CNN/Yankelovich, 2/10/94.

Consequences of Low Support

Decisionmakers typically seek to insulate their operations from political opposition, especially when public support appears low or fragile rather than robust. One way to do so is by imposing constraints on the operation, including those that were described at the beginning of this section.[9]

By imposing constraints, decisionmakers seek to reduce the risk that the situation will deteriorate in ways that will lead to active opposition; put another way, constraints are imposed to reduce the probability that the leadership will have to expend scarce political capital to preserve a permissive environment for military operations, rather than promoting other items on the agenda.

In sum, domestic constraints on coercion are perhaps best thought of as limitations that are placed on a military operation by decision-

---

[9]Other strategies are available. For example, as part of his casualty-minimization strategy on Bosnia, President Clinton sought to deter attacks on U.S. forces by dispatching fairly heavy combat forces while simultaneously threatening an immediate and decisive U.S. military response if U.S. forces were attacked. In essence, the credibility of the U.S. threat, and the success of its casualty-minimization strategy, were in large part made possible by the capabilites of the forces that were sent to Bosnia. See Larson (forthcoming).

makers to avoid undesirable political costs. As will be seen, the imposition of these constraints can provide ample opportunity for mischief or manipulation by adversaries.

SOURCES OF DOMESTIC CONSTRAINTS

There are four major sources of domestic constraints—the administration's own declaratory policies, the Congress, the media, and the public—and each domain interacts with the others.

Declaratory Policy

Policymakers often find that declaratory policy dictates or otherwise restricts the tools available for coercion, or restricts the range of concessions that can be offered to an adversary. For example, President Clinton pledged not to put U.S. troops on the ground in Bosnia until a peace agreement was in place. Until the Dayton Accords, the Clinton administration's coercive strategy relied primarily on air power, rather than the threat of dispatching ground forces, which had been ruled out by declaratory policy. Similarly, when NATO initially launched Operation Allied Force in March 1999, Clinton administration officials announced their intention to send ground forces only to a "permissive" Kosovo environment. In short, constraints may result simply from the web of restrictions and need for consistency and continuity that are reflected in extant declaratory policy, national law, or treaties and other agreements.

Departures from declaratory policy may result in widely reported administration debates over the wisdom of straying from current policy, or bureaucratic efforts to sabotage policy change through delay, selective leaks, or other actions. Changes to declaratory policy will certainly result in close scrutiny from within and without an administration, and in many cases also will result in criticism and debate from the Congress, press, or others.[10]

---

[10]Recent congressional criticism of the evolving U.S. policy toward Cuba and China comes to mind. The change of objectives (policy) in Somalia in the spring of 1993 occasioned only a small amount of criticism at that time, but attacks on U.S. "mission creep" from the objective of establishing a secure environment for humanitarian relief

Congressional Checks and Balances, and Other Restrictions

As in many diplomatic endeavors, the checks-and-balances nature of the U.S. government can restrict U.S. policymakers' room for creativity or flexibility in crafting coercive strategies. Although it has generally balked at constitutional challenges to the president's war powers,[11] the Congress nevertheless may be the most important single source of influence on the constraints a policymaker imposes on a military operation. Among the sorts of resulting constraints are:

- *U.S. Laws.* U.S. laws may impose restrictions on a president's military operation, or may require extensive consultation, explanation, or documentation.[12] Consequently, in an effort to limit political costs, presidents have typically limited forces to those that could be mobilized without call-ups of the military reserves and use of the Civil Reserve Air Fleet (CRAF). Reserve call-ups and activation of the CRAF not only constitute noisy political signals that a situation is escalating, but they also can involve dislocations to businesses, which can result in political costs.[13]

---

operations to one of internal security and nation-building were quite common by late summer 1993.

[11] For a detailed treatment of congressional unwillingness to formally press the war powers issue over Bosnia, see Hendrickson (1998). Hinckley (1994) suggests that this is typical.

[12] For example, the United Nations Participation Act of 1945 preserved, in 22 U.S.C. §287(d), a congressional role in authorizing the provision of U.S. armed forces to the United Nations in a combatant role. By comparison, the president was given somewhat greater latitude on decisions involving noncombat operations, although the Congress also imposed limits on these operations. When U.S. forces already have been dispatched, the War Powers Resolution calls for presidential-congressional consultations anytime U.S. armed forces are introduced "in numbers which substantially enlarge United States Armed Forces equipped for combat already located in a foreign nation."

[13] 10 U.S.C. §§ 12301, 12302, and 12304 provide the president with authority to order reserve units and individual members to active duty across a wide spectrum of operations, and under a variety of conditions. They include national emergencies declared by the president or Congress, and Presidential Selected Reserve Call-Up (PSRC) authority, under which a president may determine that it is necessary to augment active forces for operational missions, and where up to 200,000 Selected Reservists can be called to active duty for up to 270 days. See GAO (1997). There are three stages of incremental activation in CRAF: Stage I is for minor regional crises; Stage II is for major regional contingencies; and Stage III is for periods of national mobilization. USAF (1997). President Bush accepted the political costs of his decision to deploy large numbers of U.S. forces to the Gulf in 1990, including call-up of reserves

- *Congressionally Mandated Restrictions.* Although they may lack the force of law, there are a range of ways the Congress can exact restrictions on a military operation, including denial of congressional authorization, disapproving resolutions, and specific reporting requirements. These restrictions provide opportunities for the Congress to signal its opposition or the limits of its support and to impose political costs on the president and his senior advisers for pursuing intervention policies that deviate from congressional preferences. For example, the Congress may threaten to vote on the War Powers question, or through complaints about the absence of prior consultation, insistence on extensive consultation, on-the-record votes of disapproval, funding restrictions imposed through the budgetary process, or calls for exit strategies with "dates certain" may seek to influence the mission, force levels, conduct, or duration of a military operation.

- *Formal Processes.* Congress may dictate elaborate, formal, and public procedures to ensure that the broad parameters of administration policy are subject to congressional scrutiny and debate (e.g., annual renewal of China's MFN [most favored nation] status). The result is that the administration is forced to spend political capital in exchange for congressional acquiescence to its favored policies, while being forced to justify publicly unpopular intervention policies.[14]

The Press

The media plays a critical role in the dissemination of information about, and positions on, U.S. military operations. Despite the beguiling attractions of "CNN effects" and other "media effects," most sys-

---

and CRAF. Nevertheless, the decision to create an offensive option also resulted in extensive congressional debate and criticism, and imposed a cost on the president: it effectively precluded promoting any other (e.g., domestic) presidential agenda items until the war was concluded.

[14]For example, renewal of funding for the Bosnia operation has consistently resulted in a lively congressional debate over the progress, cost, and expected duration of the operation. While congressional unhappiness over the extended duration of the operation has not resulted in a cutoff of funding, it has forced the administration to spend a great deal of political capital on justifying its continuation, and has probably helped to ensure that no more than four in ten have supported the operation.

tematic data-based research suggests that there is less to these effects than is typically believed.[15] For example, there is rather clear evidence that the volume of news reporting on military operations and other policy issues closely follows changing objective events and conditions (e.g., the level of political and military activity), and that the tone of media reporting is in any case generally indexed to the tone of the leadership debate.[16] This is not to say that policymakers do not feel pressured by the media, or feel that their decision cycles are shortening as a consequence of the advent of CNN live news broadcasting. But strictly in terms of *content*, the media seems generally to amplify attacks from other, particularly political, quarters, rather than being the source of these attacks.[17] The most accurate characterization seems to be that the media exercises maximum influence when there is an information or policy vacuum. Put another way, when political leaders fail to provide a compelling justification and explanation for their policies, or policies are failing and leaders fail to provide a viable alternative, amplification (and even distortion) by the media may take place.

The mortar attack on the Sarajevo marketplace in February 1994 serves as an example.[18] U.S. policy had failed to bring an end to the conflict in Bosnia, and generally had proved incapable of providing security to the inhabitants of Sarajevo. As a result of the policy failure, then Secretary of State Warren Christopher—having in January been severely criticized by French Foreign Minister Alain Juppe for

---

[15]See Larson (1996a, 1996b) for detailed analyses of the "CNN effect" in Somalia; others who have looked at this question (e.g., Strobel, 1997) have confirmed these findings. The data suggest that the media actually was following U.S. political leaders' actions on Somalia, not vice versa, and that the media coverage of the dead U.S. serviceman's body being dragged through the streets of Mogadishu was not responsible for the basic preference for a U.S. withdrawal from Somalia—this preference existed well before October 1993.

[16]In other words, when most leaders and experts support a policy, the media generally uses a favorable tone in reports on that policy, and when they oppose it, media commentary reflects that negative tone. On "indexing," see Bennett and Stam (1998). On the tone of media reporting on Vietnam, see Hallin (1986). See Zaller (1992) on the relationship among leadership positions, the media, and mass attitudes, including on the Vietnam issue.

[17]That is, the media finds irresistible a good fight between a president and his critics.

[18]For a detailed discussion of the limited role of the media in the policy change after the marketplace mortaring in Sarajevo, see Strobel (1997), pp. 153–159.

U.S. policy on Bosnia—was advocating a more forceful stance. The mortar attack killed dozens of innocent civilians and led to criticisms of U.S. policy by many foreign policy analysts and pundits. The attack and subsequent criticism were covered extensively by the media, particularly the electronic news media. The coverage of the event, and the calls for stronger action, strengthened the hand of Secretary Christopher and others within the administration who already were advocating more forceful action. In this case, a tragic incident focused attention on a failing policy. The media played a role secondary to the overall policy failure and vacuum. It did, however, help Christopher and others change policy to the one they already advocated. In short, under the right circumstances, "media spectaculars" can draw attention to a failing policy and force a change, but the role of the media is subordinate to the actual events that reveal policy failures.

SOME CORRELATES OF PUBLIC SUPPORT

As suggested by Table 5, perceived benefits, prospects for success, and costs each have different faces. In fact, there are a great number of specific factors that Americans consistently deem to be quite important in the initiation and conduct of U.S. military operations, and any of which can affect the resulting level of support for a military operation.[19]

The table shows that the preeminent consideration (86 percent of those polled agreed as to its importance) is the number of American lives that might be lost; the cost in dollars is generally considered to be of much less importance. The length of time of fighting and possibility of failure (which address both costs and the prospects for success) are also critical factors, and were mentioned by over half of those surveyed.[20]

---

[19]Although these data are over ten years old, the importance of the listed factors has remained fairly constant. This question is used because it appears to be the single best summary of the factors the public considers to be important in military operations.

[20]The data seem to suggest that the two factors interact significantly. For example, there is evidence in the public opinion data that when victory is expected (e.g., in World War II or the Gulf War), longer commitments are more acceptable than in cases where an extended commitment is believed to make little difference to the ultimate

Table 5

Importance of Various Factors in Use of U.S. Armed Forces
(in percentage)

No one wants our nation to get into any conflicts in the
future, but as in the past, our leaders might someday decide
to use our armed forces in hostilities because our interests
are jeopardized. I know that this is a tough question, but
if you had to make a decision about using the American
military, how important would each of the following factors
be to you?

	Very Important
Number of American lives that might be lost	86
Number of civilians that might be killed	79
Whether American people will support	71
Involvement by major power (e.g., USSR, PRC)	69
Length of time of fighting	61
Possibility of failure	56
Whether allies/other nations will support	56
Fact that we might break international laws or treaties	55
Cost in dollars	45

SOURCE: Americans Talk Security #9, September 7–18, 1988.

The table also identifies several other notable factors :

- The number of civilians who might be killed was the second most frequently mentioned consideration, with 79 percent of those polled mentioning it as being very important

- The support of the American people was mentioned by 71 percent

- Involvement by a major power (i.e., whether dangerous escalation is likely) was considered to be very important by 69 percent of those surveyed

- Support from allies and conduct of the operation in accordance with international laws and treaties were each mentioned by more than half of those polled.

outcome (e.g., in Bosnia). As was described above, since support for military operations is associated with beliefs about the probability of success, the possibility of failure can greatly reduce support.

These data confirm that in the abstract, support for military operations results from a mixture of humanitarian, moral or legal, and pragmatic considerations—what might be thought of as a humane rationality. To the extent that decisionmakers initiate and conduct military operations within these broad parameters, the probability increases that support will be high. But to the extent that military operations can be pushed outside of these broad parameters, support may decline. And to the extent that these parameters reduce the effectiveness of U.S. military action, adversaries stand to benefit.

Also worthy of mention is the issue of using force only as a last resort: there typically is much higher support for diplomatic solutions than for military solutions, and higher support for military operations when a majority believes that all reasonable diplomatic avenues have been exhausted. The U.S. public has a preference for diplomatic solutions over the use of military force. If force is used, in general the public prefers that it be used in a multilateral setting.[21] This final preference is elaborated in the next chapter.

Some Conditions Under Which Support Is Typically High

Support for coercive military operations is likely to be high in the face of:

- *Counterterrorism.* There generally is widespread support for using air power to punish terrorist groups and their sponsors, when they can be positively identified.[22]

- *Weapons of mass destruction* (WMD). Limited evidence suggests high support for countering WMD, and although there are rational concerns about escalation to nuclear use, presidents are generally granted wide latitude in such actions. For example, 89 percent of those polled in the first half of 1998 said that Iraq would pose a serious threat if it acquired nuclear weapons,[23] and

---

[21]Gallup, January 1998, discusses this preference with respect to the crisis in Iraq.

[22]For example, the August 1998 cruise missile attacks on the terrorist facilities in Sudan and Afghanistan were supported by 75 percent of those polled by the *Los Angeles Times* on August 22, 1998. See Barabak (1998).

[23]Gallup, June 5–7, 1998.

between about two-thirds and three-quarters of those polled have consistently expressed support for air strikes to coerce Saddam to cooperate with United Nations Special Commission (UNSCOM) inspectors.[24] A plurality of 47 percent supported military action to reduce Iraq's ability to produce weapons of mass destruction even if it would result in substantial casualties among Iraqi civilians.[25] And 61 percent supported continuing the embargo on Iraq until all WMD capabilities were eliminated, even if innocent Iraqi civilians suffered as a consequence.[26] When President Clinton launched Operation Desert Fox in December 1998, about three-quarters of the public approved.[27]

- *Halting invading armies vs. internal interventions.* For a host of reasons—not the least of which seem to be the rather spotty U.S. performance record in interventions to create political order out of political-military chaos—Americans are generally much more supportive of operations that involve a U.S. contribution to the collective self-defense of a friend or ally than operations involving internal interventions in messy civil wars and other internal conflicts.[28]

There are other cases where support is highly contextual, and sensitive to the risks and costs of different courses of action:

- *Americans at risk.* The presence of Americans at risk can yield higher or lower support for the use of force, depending on the

---

[24]This is from a series of questions asked about whether respondents favored or opposed the United States using its air force to bomb Iraqi targets if the government continued to restrict UN weapons inspections. Results were as follows: CBS News, 2/1/98 (74 percent); CBS News, 2/8/98 (69 percent); CBS News, 2/17/98 (77 percent); CBS News/*New York Times*, 2/19–21/98 (66 percent); Princeton Survey Research Associates, 2/19–22/98 (67 percent); and CBS News, 3/1–2/98 (65 percent).

[25]Gallup/CNN/*USA Today*, 2/13–15/98. Forty-five percent opposed. As a point of comparison, 42 percent favored and 49 percent opposed military action to force Saddam from power that might result in substantial Iraqi casualties.

[26]AP, 1/2–5/92. In August 1990, 54 percent of those polled had supported cutting off all food to Iraq even if it meant innocent civilians would go hungry. NBC/*Wall Street Journal*, 8/18–19/90.

[27]Connelly (1998), p. A26.

[28]See Jentleson (1992) and Jentleson and Britton (1998) on this point. For a relatively recent review of U.S. public attitudes on the use of force, see Kelleher (1994).

context. Generally speaking, there typically is high support for using force to secure the release of Americans held hostage, or to punish those responsible for the deaths of Americans.[29] In other cases, the holding of hostages has not been an effective deterrent to using force, although Americans are concerned about military operations that might endanger the lives of U.S. hostages.[30] Furthermore, there has been strong support for actions, including the use of force, that would reduce the prospects for hostages being taken.[31]

- *Rescuing allies at risk.* Similar to the contextual nature of support for the use of force when Americans are at risk is support for rescuing beleaguered allies and friends in a coalition situation. There was high support for the narrow objective of *rescuing* NATO peacekeepers in Bosnia, for example, and much lower support for other types of U.S. intervention that would have entailed a deeper U.S. commitment.[32]

---

[29]During the Mayaguez crisis (1975), majorities approved the rescue operation, including the bombing of Cambodian forces, even though almost as many Marines died in the operation as Americans were rescued. During the Iran hostage crisis (1979), although there was little support for military action against Iran, majorities did approve the Iran hostage rescue effort, even though it ended in disaster. Few supported the use of force to resolve the Lebanon hostage crisis of the 1980s. Diplomacy was favored over military force in resolving the issue of Americans held hostage by Somali warlord Aideed and his clan in October 1993. Although majorities generally preferred diplomacy over force in resolving the return of American pilot Captain Scott O'Grady from Bosnia, it seems reasonable to believe that the search and rescue operations that were undertaken were supported (there are no data available on this specific question).

[30]Polling before the Gulf War shows that more than half continued to support the use of military action, even if Iraq continued to hold U.S. hostages; support for military action increased if Iraq killed any Americans. On 8/8/90, ABC News/*Washington Post* found 85 percent in favor of stronger military action if Iraq took American civilians hostage. Gallup/*Newsweek* found 61 percent supporting engaging in combat if Iraq continued to hold hostages in their 8/9–10/90 polling; 55 percent in 8/23–24/90 polling; 57 percent in 10/18–19/90 polling; and 55 percent in 11/15–16/90 polling. In the event that Americans actually were killed, support for combat rose to between 79 and 84 percent.

[31]There was, for example, support for coercive air strikes if Iraq tried to shoot down U.S. spy planes. According to ABC News' 11/11/97 poll, 38 percent supported major bombing attacks, and another 46 percent supported limited attacks.

[32]Regarding Somalia, there was a high level of support for a forceful U.N. response following the deaths of 26 Pakistani peacekeepers in June 1993.

- *Humanitarian operations.* Where the likelihood of combat and the risks and costs of humanitarian operations are perceived as quite low, there is high support for humanitarian operations.[33] Support is sensitive to casualties, however, and when casualties are expected, support may be restricted to four in ten or fewer.

The foregoing suggests that the American public is generally both humane and pragmatic, and that support for U.S. military operations is systematically associated with a number of beliefs or factors related to the cause and the conduct of the military operation, which can be summarized in terms of perceptions of benefits, prospects for success, and acceptable costs.

HOW AN ADVERSARY CAN EXPLOIT U.S. DOMESTIC POLITICS

The most obvious consequence of domestic political constraints on coercive strategies is that constraints limit practicable options. As Chapter Three showed, coercive threats are more likely to succeed when the coercer achieves escalation dominance, defeats an adversary's military strategy, or magnifies third-party threats. Political constraints remove options from the table that otherwise might contribute to any one of these factors.

Because coercion relies on manipulating an adversary's cost-benefit calculus, it is necessary to understand not only how these constraints actually operate on coercers but also how the adversary expects them to operate. This section therefore views the issues raised above through the adversary's eyes, identifying commonly used strategies to exploit U.S. domestic opposition and potential strategies that could be used in the future.

Adversaries frequently exploit U.S. domestic politics to decrease the effectiveness of coercive diplomacy.[34] There are a number of tactics

---

[33]Approximately three out of four supported the original U.S. mission in Somalia, even with the possibility of casualties. Initial support for U.S. participation in NATO air attacks against Yugoslavia (Operation Allied Force) in March 1999 was between 50 and 60 percent. *USA Today* (April 29, 1999).

[34]Some of these counter-coercive strategies are elaborated in Byman and Waxman (1999).

available to adversaries that can erode the bases of support by targeting specific factors associated with perceptions of the benefits, the prospects for success, or the actual and expected costs. It seems reasonable to suggest that clever adversaries may try many tactics in the hope that at least one will work to their advantage. Finally, many of these tactics are complementary, or blur together in ways that make the distinctions far from clear; the following paints with a broad brush the range of alternatives available to adversaries.

Diminishing the Interests

Adversaries have sometimes guaranteed the security of key U.S. interests in the hope that by diminishing the sense of threat to those interests, the United States will abstain from intervention. This is perhaps most often found in cases where civil disturbances abroad lead rebels to promise the security of Westerners or other foreigners, but it can also take place in other circumstances.[35] Security guarantees can be particularly manipulative when they are coupled with threats (e.g., that if the United States does intervene, the adversary will no longer have any responsibility for the protection of vital interests, or the continued safety of U.S. citizens).

Tarnishing the Cause and Conduct of the Operation

Just war doctrine and, to a large extent, the international law of war, distinguish between the legitimacy of the causes for initiating war (*jus ad bellum*) and the moral legitimacy of the actual conduct of a war (*jus in bello*). Because most Americans seek reassurance of the rectitude and morality of U.S. uses of force, each presents opportunities to potential adversaries. If adversaries can paint a picture that they are in fact conflict victims, either by making a case that they were not the initiators, or that they (or, particularly, their civilian populations) are in any case carrying the burden of the costs of the war, adversaries may be able to erode the moral force behind U.S. involvement.

---

[35]For example, prior to the beginning of the Gulf War, Iraq made numerous statements that oil supplies would not be interrupted, that Iraq would continue to make Kuwaiti oil available to international buyers, and that Saudi Arabia should not feel threatened.

In addition to attacking the justness of the initiation of a war (*jus ad bellum*), adversaries can attack the justness of the conduct of the war (*jus in bello*), rules for which are codified in various conventions and treaties on the international law of war. Adversaries may try to fall back on international law, treaties, or conventions (e.g., the Geneva Conventions), or international organizations (e.g., the United Nations) to raise questions about U.S. conduct in military operations.[36] Or adversaries can exploit restrictive U.S. targeting rules by placing legitimate military targets in essentially civilian areas,[37] or placing civilians at military targets.[38]

The U.S. public's desire to minimize unnecessary violence to noncombatants is not without limits or exceptions. Whereas the American public expects the U.S. military to minimize collateral damage in the conduct of its military campaigns,[39] under some circumstances the public appears to recognize the unavoidability of casualties: in the Gulf War, the public was mostly insensitive to Iraqi civilian casualties.[40] One plausible reason is that air strikes were

---

[36]In post–Desert Storm Iraq, for example, Iraqi leaders have attempted to use international law and disingenuous legalistic arguments to characterize U.S. military action as acts of aggression, the embargo on Iraq as unfair and illegal, UNSCOM as U.S.-dominated and not representative of the United Nations, and UNSCOM as confusing the important and the unimportant issues in their mandate. They similarly have sought to woo Russia and France, in part in an effort to weaken support for UN Security Council resolutions regarding Iraq.

[37]Both the North Vietnamese and Iraqi leadership sought to exploit this opportunity.

[38]For example, during the Gulf War, the Iraqi leadership placed civilians in the Al Firdos bunker, a military facility; when it was destroyed in the air war, Iraq tried to exploit the military and media possibilities. In fact, political leaders immediately placed additional restrictions on air operations to prevent a recurrence, although most members of the public seem to have been unmoved. In this particular case, it is unlikely that the Iraqi leadership deliberately put the civilians at risk, but it nevertheless exploited their deaths.

[39]In the Gulf War, 91 percent of those polled thought the U.S. military was doing all it could to keep down the number of civilian casualties in the war against Iraq. *Los Angeles Times*, 2/15–17/91. Fifty-eight percent of those polled by *Time*/CNN/Yankelovich on 4/21/94 opposed using military planes to bomb targets, such as roads and power plants, which the Serbs were using for civilian as well as military activities in Yugoslavia.

[40]Mueller (1994, p. 317) presents data from ABC News/*Washington Post* polls on 1/20/91 and 2/14/91 showing that the percentage of those polled who said that U.S. bombers should not pass up targets if Iraqi civilians might be killed in the attack ranged from 56 to 60 percent; polling by ABC News/*Washington Post* on 2/8–12/91 and 2/14/91 found that the percentage who said that the United States was making

seen as a way of minimizing U.S. deaths in ground combat, and when forced to choose between U.S. and Iraqi deaths, most sided in favor of enhancing the safety of U.S. soldiers.[41] Another reason may have to do with the degree to which Saddam's regime had been demonized, and the consequent willingness to accept less-stringent constraints. Sometimes, deaths to noncombatants can be seen to be the responsibility of the adversary, essentially backfiring.[42] For example, when the Al Firdos bunker was destroyed during the air war in the Gulf, a large number of civilians were killed, and the Iraqis argued that a civilian bomb shelter had been attacked. Most of the American public found more credible the U.S. leadership's argument that civilians had purposefully been collocated in what was essentially a military facility, and blamed the Iraqis—not the U.S. military —for these deaths.[43]

A related strategy available to adversaries is to tarnish the war or operation indirectly, by splitting off the United States from its allies or coalition partners. If differences between the United States and its allies or coalition partners can be created, whether over the objectives, strategy and conduct, forces used, or risks and costs assumed

---

enough of an effort to avoid bombing civilian areas in Iraq ranged from 60 to 67 percent.

[41]Sixty-nine percent of those polled during the Gulf War agreed with the statement that "the deaths of civilians who are located close to a military target are worth it if American lives are saved." *USA Today*, February 15, 1991.

[42]Seventy-one percent of those polled said that the United States was justified in attacking military targets that Saddam had hidden in areas populated by civilians (*Los Angeles Times*, 2/15–17/91); 67 percent said they thought the United States was making enough of an effort to avoid bombing civilian areas in Iraq, while only 13 percent said the United States should make a greater effort (ABC News/*Washington Post*, 2/14/91); and 90 percent said they did not think the U.S. military had intentionally bombed civilians in the war (ABC News/*Washington Post*, 2/14/91).

[43]Eighty-one percent of those polled thought the bunker was a legitimate military target (ABC News/*Washington Post*, 2/14/91), and only 3 percent in a 2/15–17/91 poll by the *Los Angeles Times* believed that the bunker was solely a bomb shelter for civilians; 23 percent thought it was an Iraqi military shelter, and an additional 59 percent thought that it was both an Iraqi military shelter and a civilian one, i.e., that the Iraqi leadership had collocated civilians in a legitimate military target. Nevertheless, the expected public reaction did restrain subsequent military action. The bunker incident led to severe restrictions on further strikes against Baghdad targets. Despite the desire to avoid Vietnam War–style Washington micromanagement of bombing campaigns, the incident led to a reversal of policy, after which strikes on Baghdad or other politically sensitive target required specific approval of senior decisionmakers.

by different coalition members, adversaries may find that these differences raise new questions at home about the legitimacy of the operation. The aim of an adversary is to create and exploit divisions that ultimately will result in the loss of coalition partners, particularly those who presumably have a greater stake or interest in the outcome. Adversaries will hope to raise the question: If those who are most affected by the outcome have lost interest, why should the United States continue?

Dragging Out a Conflict

As discussed above, support for a policy option is typically related to the belief that it has prospects for success. By dragging out a conflict, an adversary can often lead the United States to abandon a coercive campaign. In the same way that the perceived stakes for local actors are likely to be much higher than those for the United States, time preferences can favor adversaries and others whose interests are more directly at stake. Although this time preference is not as acute as many believe,[44] it still remains an effective basis for constructing an asymmetric strategy.[45] A strategy available to adversaries is to play a waiting game, nominally acquiescing to U.S. wishes to avoid punishment, while engaging in passive resistance to specific, tangible steps that would move the situation to the desired U.S. outcome.

The inherently political requirement that an "exit strategy" provide a "date certain" by which U.S. troops will conclude an operation is an example of a constraint that is sometimes acquiesced to by policymakers to minimize opposition to an operation.[46] This acquiescence

---

[44]For example, public opinion data suggest that most Americans expected a somewhat longer war in the Gulf than the short war that eventuated. See Mueller (1994, pp. 305–306). The data furthermore suggest that support for the war would probably have held up for several (perhaps three to six) months, so long as there was continued evidence of success. See Larson (1996a, 1996b).

[45]This is not peculiar to the United States, but may be a vulnerability of many democracies. In their analysis of interstate wars between 1816 and 1990, Bennett and Stam (1998) found that after roughly 18 months, democracies become far more likely than autocracies to quit a war, and more willing to settle for draws or losses.

[46]The issue of "exit strategy" and a "date certain" sometimes becomes confused. To be clear, there is nothing wrong with an exit strategy, i.e., a strategy that is expected to lead to a specific outcome and make possible the removal of U.S. military forces. Problems may arise, however, when an exit strategy includes a "date certain" by which

creates vulnerabilities, however, in that it exacerbates asymmetries in time preferences. Moreover, it provides adversaries with strong incentives to wait until the deadline has passed while arming and engaging in organizational or other activities that will improve their prospects in a post-U.S. environment. As shown in Bosnia, when objectives have not been achieved by the initial deadline, it can occasion a great deal of criticism from opponents and necessitate the expenditure of presidential political capital to shore up support for continuation.

Another strategy for fostering the perception of a hopeless situation is to engage in an unrelenting pattern of low-level guerrilla warfare. Sniping, mortar, or artillery attacks, or minings and small-scale terrorism can contribute to the perception that there are no "centers of gravity," and therefore little chance to improve the situation through U.S. military action. Vietnam, Lebanon, and Somalia are all good examples of cases in which large segments of the public had the sense that the United States was becoming bogged down in a quagmire where military action was incapable of improving the U.S. position. Even in Bosnia as recently as late 1997 and early 1998, questions surfaced about whether sufficient progress was being made on the political front (the avowed *sine qua non* for the withdrawal of U.S. forces) to warrant continuation of the mission. Adversaries could try to create a quagmire deliberately to force a U.S. withdrawal.

Imposing Costs

In addition to undermining perceived benefits and reducing the likelihood of success, adversaries can impose costs directly on the United States. Perhaps the most effective means of imposing costs is killing U.S. soldiers. In cases where U.S. equities are small, and where support for the operation was never robust, casualty-generating strategies will raise the profile of the operation and arouse new doubts about the aims, strategies, costs, and prospects for the operation, questions that policymakers would prefer not be raised. Judging from Lebanon and Somalia, a common response is to move forces in place to a more defensible position until an orderly with-

––––––––––––

time U.S. troops are expected to be removed. For a discussion of exit strategies where this critical distinction is not made, see Rose (1998).

drawal can be executed, usually three to six months later, while claiming that the United States is not being driven out.[47]

In situations where U.S. interests are compelling, however, the loss of lives will typically have little effect on support for a continued U.S. presence. Following the Al Khobar bombing, the focus was on determining responsibility for force protection and improving the security of U.S. personnel in Saudi Arabia; there was never serious discussion about withdrawing from Saudi Arabia as a result of the bombing. Put another way, in the minds of most leaders and members of the public, there is a significant difference between Saudi Arabia and Lebanon (or Somalia, for that matter) that conditions the response to casualties there.

In situations where U.S. interests are reasonably compelling—or when U.S. servicemen are being held hostage—adversaries who use a casualty-generation strategy risk a backlash. As described above, the tolerance for casualties is much higher in cases like Saudi Arabia than in cases like Somalia, because the stakes are generally perceived to be more important. Even in Somalia, however, the capture and mistreatment of U.S. servicemen angered many, and most Americans were willing to support strikes on Aideed if diplomacy failed to release the U.S. servicemen held by the clan leader, or to punish Aideed after they were returned. This hardening of U.S. resolve is particularly likely if adversary moves include terrorism. The high levels of support for the U.S. air strikes on Libya in 1986, and the support for Tomahawk missile strikes against Iraq after the discovery of the plot to assassinate former President Bush, suggest that most Americans support the punishment of terrorists with air strikes and cruise missiles.[48] Similarly, in Somalia the U.S. public would have

---

[47]In a sense, the preeminent objective becomes force protection, and the pursuit of additional objectives is eschewed. This arguably occurred both in Lebanon and Somalia. Taw and Vick and others have argued that force protection concerns in Bosnia significantly impede U.S. operations. Thus, even before a single casualty occurs, force protection can become a major hindrance to achieving U.S. goals. See Taw and Vick (1997), pp. 194–195.

[48]In 1981, only 10 percent favored sending American planes to bomb terrorist training camps in Libya. Harris poll, 12/13/81. By 1986, with the disco bombing attributed to Libya, support was much higher: 47 percent supported limited air strikes and 20 percent supported an extensive military invasion (NBC/*Wall Street Journal*, 4/13–15/86); 56 percent approved strongly and 21 percent approved somewhat of the U.S.

supported strikes against Aideed after U.S. forces were out of harm's way.

Adversaries can also impose costs by targeting friends and allies. Multilateral operations typically receive higher levels of support than unilateral ones.[49] Adversaries can selectively attack the forces of U.S. friends, allies, or coalition partners in an effort to split off coalition members from the United States, searching for the "weak link" in the coalition. This strategy can be used to test the will of the least committed in an effort to bring pressure to bear, encourage them to press for a softer strategy, or perhaps even to drop out of the coalition. This issue is further elaborated in the following chapter.

Manipulating Asymmetries in Escalation Preferences

In addition to manipulating public views of the benefits of intervention or generating public opposition to a military operation, adversaries may seek to exploit inherent asymmetries in the ability or willingness to escalate to higher levels of violence. Adversaries thus may raise the level of violence to exploit escalation advantages.

Adversaries may see brinkmanship as a productive strategy for demonstrating the ineffectiveness of threats or use of military force, including air power. Saddam has routinely and regularly created crises to challenge the United States and, to a lesser extent, the United Nations, over the mandate and access of UNSCOM weapons inspectors. After a number of years of this sort of brinkmanship, large percentages believe that military action restricted to punishing

---

air strike against Libya (ABC News/*Washington Post*, 4/24–28/86); and 75 percent approved of the military air strike against Libya (ABC News/*Washington Post*, 5/15–19/86).

[49]Indications of less-than-widespread support from coalition partners can sap support. For example, where approximately two-thirds of those polled by CBS News/*New York Times* consistently supported U.S. Air Force strikes against Iraq to force the Iraqis to lift restrictions on U.N. weapons inspectors, an alternative question suggesting United Kingdom participation in joint action with the United States but opposition from some allies like France and Russia yielded support from much smaller percentages: 66 versus 58 percent (2/19–21/98), 77 versus 62 percent (2/17/98), and 69 versus 53 percent (2/8/98).

Iraq over the weapons of mass destruction issue will be insufficient to change Iraqi behavior. Through continued intransigence, the Iraqis hope to demonstrate the ineffectiveness of coercive diplomacy and air power, and increase pressures on the administration to undertake escalation that will be politically costly. Saddam Hussein may have provoked the December 1998 confrontation, which resulted in four days of U.S. and British cruise missile and air attacks (Operation Desert Fox), calculating that his regime could better withstand the political costs of new levels of violence than could the United States and its coalition partners.

As discussed earlier, both the American public and decisionmakers are sensitive to the risk of U.S. casualties. Adversaries can exploit escalation fears simply by threatening to raise or actually raising the level of violence to find the level at which U.S. domestic support turns to opposition.[50] Except in cases where vital, even existential, interests are engaged, large segments of the public seek to avoid situations where the consequences of escalation are profound (e.g., threats of Russian or Chinese involvement, or a nuclear exchange) and might result in risks that are incommensurate with the stakes at hand.[51] This means that in some cases (e.g., Iraq, North Korea, China, and Taiwan), majorities of the public have supported a strong U.S. military response to adversaries who are engaging in a game of escalation. It also means that in other cases, where the stakes are perceived to be small, and/or the United States seems to be in a losing situation (e.g., Lebanon, Somalia), there is typically little interest in escalating or increasing the level of commitment beyond what is required to protect forces in place, or to recover U.S. servicemen

---

[50]In the Vietnam War, for example, only 25 percent of those polled in December 1967 favored an escalation of the war that would involve bombing supply lines and air fields in China. (Harris poll, 12/67). In the Gulf War, between 16 and 25 percent favored Israeli retaliation for Iraqi missile attacks (which would have resulted in a widened war and complicated U.S. relations with Arab coalition members), whereas a majority consistently supported U.S. retaliation.

[51]Nearly seven in ten mentioned the involvement of other actors such as Russia and China as being very important in a military operation.

held hostage.[52]   In such cases, an orderly withdrawal is typically preferred over escalation.[53]

Horizontal escalation—fostering an expansion of the conflict across provincial or national borders, for example, or drawing in other actors who will complicate U.S. calculations—also may be a desirable strategy for adversaries, particularly where such expansion is feared. Adversaries have regularly used sanctuaries in other countries to complicate U.S. operations, effectively daring the United States to strike the sanctuaries and create unwanted escalation. North Korea received support from Chinese sanctuaries during the Korean War; North Vietnam challenged U.S. actions during the Vietnam War by operating from Laos and Cambodia. These efforts made it far more difficult for the United States to coerce without raising undesired costs for itself.

CONCLUSIONS

The data presented here suggest that the level and robustness of domestic support are highly contextual in nature, and depend greatly on the merits (perceived benefits, prospects, and costs) in any given situation, and that presidents will continue to impose constraints on U.S. military operations to ensure the political viability of these operations. In the current domestic environment, most U.S. military operations for the foreseeable future probably will be undertaken with mixed or less-than-majority support. Decisionmakers accordingly will impose constraints on these operations to hedge against the possibility that military actions will turn into political liabilities, and under many circumstances these constraints may in turn reduce

---

[52]In Somalia, for example, Aideed raised the level of violence past the point at which the United States and its coalition partners were willing to make additional sacrifices for what had originally been a humanitarian operation. The Clinton administration reportedly considered sending armored personnel carriers (APCs) to Somalia in the summer or 1993, but reconsidered, possibly because of concerns that such an action would be perceived by congressional critics as tangible evidence of "mission creep" and an escalation of the U.S. commitment in Somalia. In the end, the administration was criticized anyway: the failure to send the additional forces was cited by critics as evidence of the administration's failure to provide adequate force protection.

[53]"Orderly withdrawal" means a withdrawal following the recovery of dead, wounded, and captured, taking place over three to six months to avoid the impression that casualties inexorably will lead to immediate withdrawals of U.S. forces.

the effectiveness of U.S. forces. The principal challenge for the Air Force may be to develop air power capabilities that can create viable military options in these situations.

Air Force planners must develop the capabilities that can effectively achieve U.S. objectives within stringent constraints, or make these constraints entirely irrelevant. Put another way, to be a useful and effective coercive instrument, air power must be able to accomplish objectives at lower political costs than alternatives could do and also be able to accomplish objectives within the likely imposed constraints.

Chapter Five

COERCION AND COALITIONS

The United States will frequently conduct coercive operations as part of a multinational coalition.[1] Crises in Kosovo, Bosnia, Somalia, and Iraq are only the latest instances where coalition unity (or the lack thereof) proved a key factor in the success or failure of coercion. Coalition partners may bring military, diplomatic, or other forms of support, but coalition-building and maintenance can also undermine coercive threats and offer adversaries counter-coercive options.

With a few exceptions, all coercive military operations carried out by U.S. forces since the end of the Second World War have been prosecuted under the auspices of international organizations or ad hoc collections of interested states. Coalitions offer many advantages to coercers. Some coercive instruments are naturally suited for broad coalitions. Coalition-building plays to the strength of economic sanctions, for instance, which require wide international support to apply pressure on target states. Coalition-building can enhance domestic public support for military operations by lending them added legitimacy. Coalition partners may also bring assets to the table that prove useful for coercive operations, including additional bases, local

---

[1] A "coalition" is a collection of actors cooperating to achieve a common objective. Coalitions include standing bodies (such as NATO) acting as a unit and also ad hoc collections of states and other international bodies working together toward a particular goal. Although the shape and size of coalitions differ widely, their key attribute with respect to coercive operations is the members' pooling of military, economic, or diplomatic efforts against a common adversary.

access, and diplomatic support.[2] When coalitions are united, coercers are better able to sustain military operations, to defeat enemy
forces, and to gather intelligence.

The picture is not completely rosy, however. Squabbling among
coalition members may result in inefficient decisionmaking and reduce the ability of the coalition to sustain and escalate military operations. The demands issued by the coalition are often watered
down to preserve unity, reducing the concessions gained. Rules of
engagement (ROE) become burdensome, increasing rather than reducing the tension between political goals and military options.
Coalition credibility may suffer as a result of bickering, making the
adversary more resistant. The adversary even gains ways to offset or
counter coercive threats if it can further divide the coalition through
escalation or half-hearted concessions. Such problems are particularly relevant for air power if they negate the speed and flexibility that
make air power such a potent coercive tool.[3]

This chapter begins by briefly noting the advantages that coalitions
offer coercers. It then discusses how coalitions can limit or hinder
attempts at coercion.

THE ADVANTAGES OF COALITIONS FOR COERCERS

Coalitions offer several potential advantages over unilateral coercive
action. For the U.S. Air Force (USAF) today, coalition contributions
in the form of bases and access are often far more important than direct contributions given the USAF's qualitative superiority over other
air forces in stealth, command and control, and precision strike.[4] As

---

[2]Riscassi (1993), and Claude (1995), pp. 49–50.

[3]These dilemmas are discussed in Byman and Waxman (1999) and Waxman (1997a).

[4]Coalition partners may contribute additional military forces or assets. When
properly employed, these assets improve the coalition's ability to sustain operations
and escalate if necessary, pose a greater chance of stalemating or defeating an enemy
on the battlefield, and improve overall intelligence collection. In Operations Desert
Shield/Desert Storm, close to 40 nations contributed military forces. In the UN
Operation in Somalia (UNOSOM II), close to 30 nations sent peacekeeping troops.
Given the growing USAF and U.S. superiority over even NATO, however, the military
value of these forces may be minimal.

noted in Chapter Four, coalition military forces also play a valuable political role, bolstering U.S. domestic support for a deployment.

Coalition partners may offer basing or overflight access that allows the United States and its allies to maximize the effectiveness of their air power assets. Although broad international support allowed for massive buildup of U.S. regional air power in 1990–1991, the erosion of this support by 1998 severely diminished U.S. and allied coercive air power potential. During the February 1998 inspection crisis, the United States had access to only 38 land-based aircraft for strikes on Iraq, because the 198 aircraft deployed in Turkey, Bahrain, and Saudi Arabia were not available in the face of host nation concerns.[5] Deployment of carrier-based aircraft to the region helped make up the lost combat potential, although aggregate firepower was substantially reduced and some specialized capabilities (e.g., F-117, Airborne Warning and Control System [AWACS], and Joint Surveillance Target Attack Radar System [JSTARS]) only operate from land bases.

Aside from yielding direct military contributions, coalition-building may produce political benefits at home and abroad that improve coercive effectiveness. The wider a coalition, the greater perceived legitimacy a military operation receives among both domestic and international audiences; conducting military operations under the auspices of international organizations such as the UN or Organization of American States (OAS) further bolsters claims that such operations accord with international norms. Even if states do not participate militarily in operations, they may be part of a broader effort to isolate the target state, and their diplomatic support may increase the credibility of threatened strikes by demonstrating international resolve. Domestic and international support for coercive operations increases the politically acceptable level of force that can be threatened and bolsters the credibility of threats by reducing the possibility of a domestic backlash.

---

[5]Defense News (1998), p. 3.

LIMITS IMPOSED BY COALITIONS

The benefits of coalitions, however, may be offset by greater implementation problems and new vulnerabilities. Coalition members typically have diverse goals or different preferences, leading the coalition as a whole to adopt positions that reflect the "lowest common denominator" rather than more assertive positions. In addition, the shared control inherent to most coalitions often makes the coalition cumbersome and its response weak, reducing its credibility. The differing goals of coalition members lower the coalition's credibility and may make sustaining operations harder. Escalation in particular becomes more difficult under burdensome ROE that coalitions often employ.

Lack of a Common Agenda

Coalition coercive operations will be complicated by an inevitable lack of harmony among members' interests. As Alexander George and William Simons speculated:

> Coercive diplomacy is likely to be more difficult to carry out when it is employed by a coalition of states rather than by a single government. Although a coalition brings international pressure to bear on the target of diplomacy and can devote greater resources to the task, the unity and sense of purpose of a coalition may be fragile.[6]

This prediction has been borne out by recent crises and international responses. In the aftermath of the Gulf War, economic motivations reduced French and Russian support for a confrontation, as they sought to renew trade and financial relations with Iraq. In Somalia, Italian resistance to military assaults reflected a preference for reconciliation with, rather than the marginalization of, Aideed's faction. Grand strategy, differing threat perceptions, third party relations, and a host of other influences caused members' policy preferences to deviate, despite initial commitment to a common goal.

Even when coalition members share a common ultimate goal with respect to the adversary, their interests are unlikely to be perfectly

---

[6]George and Simons (1994), p. 273.

aligned. The Clausewitzian notion that military operations reflect political purposes suggests that true coalition unanimity with regard to the application of force requires harmony of ends. But each state brings to the table its own strategic and political interests. Each state's military will be guided by its own set of doctrines and preferences for certain military instruments. During the course of coercive operations, individual members' interests may further diverge as a result of contingencies or asymmetries of vulnerability. For instance, if a particular contingent suffers disproportionate casualties, a member may seek a change in policy to limit further harm, contrary to the preferences of other coalition members.

The international response to the Bosnian conflict illustrates how coalition objectives may begin to diverge, even if the members ostensibly share a common objective with regard to the targeted adversary. The conflict implicated peripheral U.S. security concerns, while European countries were reluctant to get drawn into yet another Balkan conflict. A variety of factors combined to limit the potency of threats available to the allied coalition, despite an abundance of air assets at its disposal. U.S. and European operational perspectives diverged. Only the European partners had UN Protection Forces (UNPROFOR) troops on the ground. Whereas the United States generally favored a more robust coercive air strategy, Britain and France resisted, fearing that air strikes would provoke retaliatory responses by the Serbs against vulnerable coalition ground personnel. Russia, bound by traditional ties to the Serbs, resisted heavy-handed approaches. It was not until the summer of 1995, just prior to Operation Deliberate Force, that coalition members' objectives and preferences converged sufficiently to allow for robust air strikes.

Similar problems plagued attempts to coerce Saddam. In January 1993, the United States shot down several Iraqi aircraft and launched air strikes against military targets in response to Iraqi incursions and deployment of antiaircraft missiles in protected zones. The resulting widespread opposition to U.S. military action among coalition partners gave rise to speculation that Saddam had deliberately incited U.S. reprisals to win Arab support for the lifting of sanctions.[7]

---

[7]Fineman (1993), p. A1. As one example of coalition resistance to U.S. strikes, a former Egyptian ambassador to the United States urged "a pause from the policy of military escalation against Iraq in order to stop the rapid erosion of favorable Arab

Turkey, which provided key air bases supporting no-fly zone enforcement, worried that an extended conflict could contribute to its own crisis involving separatist Kurds. Russia, under pressure from nationalist hard-liners eager to reestablish economic ties with Iraq, criticized U.S. air strikes as inconsistent with international law and unauthorized by the UN Security Council. Arab states, fearing public backlash in response to U.S. military action against a regional power, urged Washington to call off further strikes.[8] Vocal criticism from Gulf War partners may have emboldened Saddam Hussein to test the coalition's resolve in 1996, 1997, and 1998 and convinced him that provocation might be an effective strategy for breaking international efforts to isolate him.

The erosion of coalition support for coercive air strikes to rein in Iraq was further exposed during the crisis precipitated by Iraq's expulsion of weapons inspectors in late 1997. With the exception of Kuwait, no Arab nation endorsed American military threats in early 1998.[9] Although diplomatic efforts, aided by the threat of U.S. military action, resolved the crisis in February, the United States found itself having to choose between employing coercive force and maintaining coalition support. Military operations were more difficult because regional allies did not want the USAF flying strike sorties out of their bases. Whereas during Desert Storm/Desert Shield coalition cohesion itself represented a key asset and may have enhanced coercive threats, the divergence of member interests several years later revealed a tradeoff between coalition unity and coercive potency.

Operations in Somalia suffered from a similar lack of unity. In launching a series of attacks to compel compliance with UN disarmament and political rehabilitation programs, the U.S.-led air attacks overstepped the bounds not only of what military force might have been able to accomplish in the Somali environment but also what the coalition members were willing to accept. From the start, the Italian contingent opposed this strong-handed approach, favoring instead a more neutral role for the UN in support of negotiations

---

public opinion which was the base of support for allied action against Saddam Hussein in the Gulf War." El Reedy (1993), p. 6.

[8]Brown (1993), p. A17; Robinson (1993), p. A25; and Wright (1993), p. A10.

[9]Jehl (1998), p. A6.

and humanitarian relief. Following the gunship strikes, Italy publicly protested the U.S.-led actions and threatened to withdraw its 2400 troops. Ireland, the Vatican, the Organization of African Unity, and several nongovernmental organizations involved in the relief effort also questioned the escalatory response. As Italy and other members condemned the anti-Aideed stance, the UN leadership asked Italy to remove the commander of its peacekeeping contingent, General Bruno Loi, charging him with refusal to obey orders from the overall UN military commander in Somalia. Loi and other members of the Italian contingent had been negotiating unilaterally with supporters of Aideed while the UN tried to isolate the Somali warlord.[10]

Even if coalition members share a common coercive objective—that is, they seek the same adversary behavior—they may have different negative objectives that limit the means by which the coalition can influence the adversary. Arab support for military action against Iraq in early 1998 waned, despite a common desire to contain Iraqi WMD production, because Arab states could not support coercive measures likely to incite domestic opposition. British and French worries of provoking Serb reprisals against their UNPROFOR contingents substantially diluted their nations' support for air strikes, even though they may have supported coercive strategies in principle.[11] In the latter case, European concern for troop vulnerabilities negated one of air power's key attributes: its ability to strike at an adversary without exposing U.S. forces to the dangers inherent in ground operations. So long as air strikes were contemplated as part of a broader coalition design that included peacekeeping forces, U.S. decisionmaking had to account for potential risks to other members' interests in fashioning a coercive strategy.

These divergent negative objectives may make it difficult to magnify a third-party threat to an adversary. After Desert Storm, for example, Saudi Arabia and Kuwait opposed the strengthening of a strong Shi'a insurgency in Iraq, fearing that Iran would expand its influence. Ankara similarly feared that any Kurdish insurgency might spread unrest into Turkey itself. Not only must coalition members agree on their objectives toward an adversary, they must also recognize that

---

[10]Lorch (1993), p. A8.

[11]Bertram (1995–1996), p. 74.

some effective means of influence may be off limits because of the concerns of one member.

Even if coalition members agree on objectives, they may differ on strategy. Thus, they may pursue inconsistent strategies or not provide sufficient resources for any of the members' preferred means to success.

The problems for coercion are exacerbated by the fact that coalitions rarely speak with a single voice—each member may at the same time communicate threats or signal messages to the adversary, perhaps in conflicting ways. As noted earlier, coercion is more likely to fail when the adversary doubts the coercer's intentions. "Clarity and consistency in what is demanded help persuade the adversary of the coercing power's strength of purpose."[12] Conflicting signals emanating from various coalition members can not only contribute to such doubts but may encourage the adversary to comply with some members' demands but not others. A possible solution is for the coalition to issue a single, common threat. This creates a lowest-common-denominator effect: to garner the necessary consensus, the coalition will gravitate toward the most restrained members' preferences. Note, for example, the January 1994 NATO summit. To paper over differences among the allies, the NATO declaration contained only vague threats of force and singled out only Srebrenica and Tuzla from among the safe areas for explicit protection.[13] The result was continued Serb bombardment of other safe areas such as Sarajevo, culminating the following month with the February marketplace explosion.

Finally, coalition members may prefer different points along the same linear spectrum. Consider several such points along a spectrum for the Iraqi crisis in February 1991:

1. Iraq remains intransigent.
2. Iraq agrees to withdraw with minimal concessions.
3. Iraq agrees to withdraw with significant concessions.
4. Iraq agrees to withdraw unconditionally.

---

[12]George and Simons (1994), p. 280.

[13]Leurdijk (1994), pp. 50–51.

U.S. efforts to disarm Iraq were nearly derailed because the Soviet Union, while sharing the common desire to see Iraq withdraw, momentarily appeared willing to accept (2), rather than support military action to achieve (3) or (4). A Moscow-sponsored initiative was eventually rejected by the United States and other coalition members, though not without threatening to draw away some members' support for military action.[14]

Shared Control

Because both positive and negative interests are rarely in perfect harmony, coalition members generally seek a say in decisionmaking to ensure that their objectives are protected. If it chooses to take the lead in forging a coalition, the United States must yield a certain degree of control over the conduct of coercive operations as the price of members' commitment to the coalition. Shared control can offset the difficulties caused by divergent interests, but it creates problems of its own. Shared control reduces the coercer's flexibility, makes escalation dominance more difficult to attain, and damages credibility.

The retained control by member states can take a variety of forms, although usually it will involve a combination of two major types: decisionmaking input or predecisional agreements. The first type, direct decisionmaking input, refers to the ability of individual members to shape case-by-case coalition decisions as they arise. In UN operations, the highest military command positions are often filled by representatives of several member nations, ensuring multinational input. Multinational political bodies generally retain some control or veto power over strategic or operational-level military decisionmaking. The result may be a dual- or multiple-key command procedure whereby the chain of command is split to ensure that all military decisions are approved by various leadership bodies.

In the Bosnia case, the United States and its allies made burdensome command arrangements to ensure that force was threatened and applied only up to a level that the coalition, as a whole, could support. To ensure that air strikes (both close air support for endangered UNPROFOR troops and air strikes against forces violating UN-

---

[14]Hoffman and Devroy (1991), p. A1.

protective resolutions) would reflect unanimous coalition support, a dual-key command and control structure governed requests.

As Figure 3 indicates, both chains of command had to approve air strikes before they could be launched. Because operational control of NATO's tactical air assets had been delegated down to the director of the Combined Air Operations Center, little coordination within the NATO chain was necessary to approve air strikes. The UN chain was burdensome, however, leading to frequent delays. UN commanders eventually agreed to delegate approval authority to lower levels, such as the UN Force Commander, although for most of the conflict, air strikes required approval by the Secretary General's special representative.[15]

The second type of retained national control, predecisional agreements, refers to arrangements negotiated in advance of contingencies to limit the actions of coalition forces. Like direct decisionmaking input, these measures help ensure that the coalition will not act contrary to individual members' interests. But rather than having to deliberate on a contingency-by-contingency basis, agreements limit coalition options to those that, ex ante, are amenable to all members given foreseeable scenarios. These agreements can increase the speed with which air power or other forceful instruments are applied, but decrease their flexibility.

At the strategic level, predecisional agreements include legal compacts, such as UN mandates, that might limit the use of force to certain levels or situations (e.g., authorizing coalition reprisals only in response to attacks on designated "safe areas" or a refusal to authorize "all means necessary"). At the tactical level, rules of engagement can restrict the freedom of coalition forces to take certain actions. Negotiating ROE provides a means by which countries providing forces can shape the conduct of coercive operations. Both legal mandates and ROE help alleviate coalition members' anxieties about unwanted escalation. The tradeoff, of course, is that they place ceilings on the potential level of force and yield escalatory initiative to the adversary, who can now more effectively dictate the level of force

---

[15]Hunt (1996), pp. 56–57.

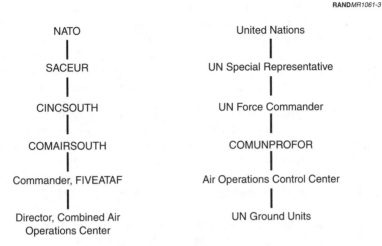

RAND*MR1061-3*

Figure 3—Command Structures in Bosnia

to be used. Given that ambiguity is an inherent part of both politics and war, compacts, mandates, and ROE often limit the effectiveness of coercion.

Most coalition operations will involve both direct input and predecisional agreements. States are unlikely to turn national forces entirely over to foreign command. At the same time, the costs of reaching consensus on every individual contingency as each arises are prohibitively high, providing the impetus to negotiate mutually amenable advance agreements on the use of force.

The two types of retained control can severely restrict coercive operations. Members' demands for decisionmaking input often conflict with demands for centralized command. Predecisional agreements, while useful for guaranteeing member support, restrict the potency and flexibility of coercive instruments. The flouting of the no-fly zone over Bosnia ultimately produced Security Council Resolution 816, which authorized active measures by NATO to control flights, and NATO began Operation Deny Flight. However, divergent demands of major coalition members resulted in a compromise agreement stipulating severe restrictions on NATO's use of force: NATO aircraft could not strike preemptively at airfields or pursue violating aircraft into Serbian airspace. The most forceful NATO response as part of Operation Deny Flight occurred in November 1994, when Serb forces launched air strikes from the Ubdina air base in Croatia to support ground operations in Bihac. In designing a response, however, NATO planners bowed to UN demands for restraint and avoided hitting Serb planes at the airfield.

ROE are designed to prevent the use of excessive force while minimizing unnecessary passivity—a balance that coalition dynamics often upset. ROE seek to authorize sufficient force so that commanders can respond to threats without appearing weak, thereby inviting further attack (Type 1 errors). They also seek to prevent commanders from using excessive force, escalating beyond political objectives (Type 2 errors). There is, of course, a tradeoff between the two types of errors, and ROE are generally formulated to balance the two as well as possible.[16] Effective sets of ROE can be thought of as the intersection of those preventing Type 1 errors and those preventing Type 2—restrictive enough to prevent escalation but liberal enough to prevent the appearance of weakness. The tension between Type 1 and Type 2 error prevention is exacerbated in the coalition context. Each coalition partner brings its own concerns regarding possible contingencies. Even if they share the same ultimate objective with respect to a common adversary, individual members may have differing negative objectives, which can produce a greater concern with Type 2 errors than would the individual members acting alone. As a result, the possible set of effective ROE may narrow to the point where there is no intersection at all.

The difficulties of balancing Type 1 and Type 2 errors in coalition contexts surfaced in the UN-issued ROE governing UNPROFOR ground operations. These orders required, among other things:

- Specific approval for offensive operations
- Minimum force necessary
- No retaliation; use of weapons only as a last resort
- Ceasing fire when an opponent ceases fire.

---

[16]Sagan (1991), p. 88.

Note that these rules allowed the adversary to avoid unintended escalation and to test the threshold of UNPROFOR military response.[17] The rigid limits on force in this case stemmed from the particular demands of peacekeeping operations, but also from the need to secure consensus among coalition members.[18]

In essence, shared control can lead to the antitheses of effective air power use: decentralized command and centralized (micro-managed) execution. USAF doctrine stresses unity of command as a key tenet.[19] However, as Air Marshal Tony Mason remarked:

> [A]mong the characteristics of air power are rapid responsiveness and high speed. It would be difficult to imagine a command and control structure more unwieldy, obstructive and operationally irrelevant to the needs of close air support than those procedures approved by the NATO council.[20]

Although the UN and NATO eventually streamlined their authorization procedures, these problems plagued coalition air operations throughout the Bosnian conflict. By the time air strikes were authorized, Serb forces had often already achieved their immediate objectives, or the moment had passed to provide endangered UN personnel with close air support.

Limits to Escalation

Escalation dominance is a key positive factor in successful coercion, but coalition dynamics often restrict the ability of the coercers to es-

---

[17]The same problems plagued ROE governing close air support for endangered UNPROFOR personnel. ROE required that hostile forces still be engaged by the time NATO aircraft arrived. On a number of occasions this allowed Serb forces to harass peacekeeping forces and then to pull back before NATO aircraft could respond.

[18]Berkowitz (1994), pp. 635–646.

[19]"Unity of command ensures the concentration of effort for every objective under one responsible commander. This principle emphasizes that all efforts should be directed and coordinated toward a common objective. . . . Unity of command is important for all forces, but it is vital in employing air and space forces. Air and space power is the product of multiple capabilities, and centralized command and control is *essential* to effectively fuse these capabilities." Air Force Basic Doctrine, AFDD 1, September 1997, pp. 12–13. (Emphasis in original.)

[20]Mason (1994), p. 177.

calate. Divergent interests—particularly with regard to negative objectives—and shared control combine to circumscribe the use of
force. ROE and legal mandates may further interfere with escalation
by placing explicit limits on practicable force.

That coalition ROE are generally approved by consensus also makes
them inflexible, in that once the parameters are set, it becomes difficult to expand them when contingencies arise. At the strategic level,
UN Security Council mandates are subject to permanent member
vetoes, meaning that such members can prevent additional autho-
rizations of force, even if they supported initial, lower-level autho-
rizations. Although the U.S. political and military leadership can re-
calibrate ROE during the course of a crisis or conflict as circumstances change, coordinating coalition ROE is a far more laborious
process. This inflexibility undermines the credible threat of escalation, often a key determinant of the success or failure of coercion.
Similar limits on escalation appear in coalition command and control schemes. A constant danger is that, in an effort to alleviate
coalition members' concerns, the organization as a whole will adopt
command and control arrangements, such as the dual-key arrange-
ment employed by NATO and the UN in the former Yugoslavia during Deny Flight, that negate the effectiveness of air power as a potent
escalatory option.

Coalition concerns also affect campaign planners' target choices, because planners must anticipate partners' political worries and protect allies from an adversary's escalation. During Operation Desert
Storm, for example, the United States limited attacks on Baghdad in
part because of the real and anticipated complaints of ostensible
allies, such as the Soviet Union and other UN Security Council
members.[21] Even more important than this restraint was the need to
use massive sorties for the "Scud hunt." The United States feared
that a failure to show Israel that the coalition was expending
considerable resources on finding and destroying Iraqi Scuds would
lead Israel to enter the war itself. Fearing that Israeli intervention
would drive Arab partners to withdraw support, planners directed
some 1500 sorties against Scud targets, roughly 3.6 percent of the

---

[21]Arkin (1997), p. 12.

total number of sorties flown.[22] Although in this case the coalition had abundant resources, in future operations the need to divert resources to operations designed to protect or placate coalition partners may trade off with options for escalation.

Finally, a coercer that chooses to operate through a coalition will often find itself faced with a choice between maintaining coalition support and resorting to new, higher levels of force.[23] In the Korean War, U.S. planners felt strong allied pressure to avoid destroying certain targets or resorting to atomic weapons use, which Britain feared might prompt Soviet reprisals.[24] Likewise, the United States could likely not have loosened the ROE for forces in Somalia following the Mogadishu firefight in summer 1993 without permanently alienating certain key coalition members. Due to the sensitivities of Arab partners (including those not participating in the strikes themselves), Operation Desert Fox planners circumscribed Iraqi targets posing high risks of collateral damage and cut off strikes after four days to avoid bombing during the Muslim holy month of Ramadan. Likewise, the particular sensitivities of each of the 19 members limited targeting options available to NATO planners during early phases of Operation Allied Force, beginning in March 1999. Even the ultimately successful U.S. strikes against Serb targets in 1995 had engendered criticism from U.S. NATO allies, which decried unauthorized escalation.[25]

---

[22]Keaney and Cohen (1993), pp. 83–84.

[23]Ironically, coalition disagreements over how to respond to Iraq's refusal to admit American weapons inspectors in November 1997 seemed to put *upward* pressure on the potential level of coercive force. While most Arab nations publicly condemned the U.S. stance of military threats, their governments intimated that, while they could not support limited, punitive strikes, they would welcome more robust strikes that incapacitated Saddam Hussein's regime. Lancaster (1997), p. A35.

[24]Futrell (1961), p. 453; Bundy (1988), pp. 242–243. A 1953 U.S. State Department report analyzing atomic options concluded that the United States "would be faced with choosing directly between Allied and neutral support and the pursuit of the proposed course of action." *Foreign Relations of the United States 1952–1954* (1984), p. 1140.

[25]Holbrooke (1998), p. 143.

Reduced Credibility

Because coercion relies on manipulating an adversary's cost-benefit calculus, the credibility of coalition threats becomes vital to favorable outcomes. One might suppose that coalition-building would enhance coercive threats in this regard, by reducing the likelihood that the coercer would back down for fear of diplomatic backlash. Somewhat counterintuitively, however, coalition threats are often less credible than those of states acting alone because adversaries recognize that members' interests diverge and see the lumbering coalition decisionmaking process as proof. Even if an adversary is wrong in its assessment, its perception of coalition disunity may cause it to hold out.

Fluctuating levels of coalition unity and national support among coalition members for military action resulted in a recurrent pattern of Serb violations of NATO ultimata during the Yugoslav conflict. In summer 1993, Serb forces were on the brink of capturing Mount Igman, the last high ground surrounding Sarajevo held by the Bosnian Muslims. On August 2, the NATO allies threatened Serb forces with air strikes unless they ceased interference with humanitarian relief to the city. Although Serb forces pulled back in the face of NATO threats, strangulation of the city resumed and culminated in the February 1994 marketplace massacre, which renewed coalition support for forceful action. Again, Serb forces pulled back in the face of a coalition ultimatum, although this time with the help of a Russian-brokered bargain. Later that year, emphasis shifted to Gorazde, where Serb forces mounted an assault on the town despite the threat of air strikes. Although NATO close air support was called in ostensibly to protect UNPROFOR troops, a rift emerged between NATO and the UN, when the UN refused to "turn its key" for further air strikes even though the Serbs had not completely complied with allied demands.[26] The eroding credibility of UN-NATO threats precipitated a crisis in summer 1995. Serb forces overran the safe area of Srebrenica in July 1995, despite a Dutch peacekeeping contingent and pinprick air strikes. Complicated decisionmaking procedures contributed to delays in providing adequate air support to Dutch units. The Zepa safe area fell shortly after. NATO threats

---

[26]Leurdijk (1994), pp. 41–66.

made the Serbs hesitate, but the lack of NATO credibility in responding quickly to Serbian aggression led the Serbs to press on with their attacks.

When coercion is conducted through a coalition, success or failure is likely to be a function of coalition unity, because a unified coalition will be able to withstand the application of higher levels of force and will also likely issue more-credible threats. But coercion is not a static process that occurs only at a single instant. It occurs over time—the United States and its partners used the threat of air strikes against the Bosnian Serbs, to varying degrees, over the course of several years. During that time, coalition unity itself is likely to vary, with important implications for coercion. Following the February 1994 Sarajevo shelling attack, NATO members achieved sufficient unity to issue a credible threat of air strikes, leading to the withdrawal of Serb artillery. This outcome contrasts sharply with the previous months' events, when differences of opinion among NATO partners surfaced at the January NATO summit, leading to intensified Serb strangulation of the city.[27]

Not only does coalition unity alter the efficacy of coercive threats, but coercive threats and strikes can dramatically affect coalition unity. The result can resemble a feedback cycle: A round of air strikes that alters adversary behavior in a positive way may rally coalition support, in turn making the threat of future rounds of strikes more credible, thereby further altering adversary behavior in a positive direction. On the other hand, a round of strikes that causes negative results (for instance, one that prompts reprisals), may cause coalition rifts, further emboldening the adversary, and so on.

The latter cycle aptly describes events in Somalia. In summer 1993, following U.S. helicopter assaults on Aideed's strongholds, Italy's public objections and threats to withdraw undermined overall UN efforts to pressure the warlord. General Aideed exploited this rift through small-scale attacks on UN personnel, thereby feeding Italian opposition against what it saw as an escalating conflict. Confrontations continued between heavily armed UN forces and Aideed's militiamen, and the use of armed force increased anti-

---

[27]Jehl (1994), p. A1; Apple (1994), p. A1; Leurdijk (1994), pp. 50–51.

foreigner sentiment among the local populace (following the July 12 raid, angry mobs killed four foreign journalists). In October, U.S. forces tried unsuccessfully to capture Aideed. During the ensuing firefight, several U.S. Blackhawk helicopters were downed and 18 U.S. servicemen killed. The UN Security Council revised UNOSOM's mandate and aborted further UN intervention in interclan conflicts, including disarmament. Soon after, President Clinton called for withdrawal of U.S. forces. Whereas coercive air strikes are supposed to alter an adversary's cost-benefit calculus by demonstrating the credible threat of increasingly painful costs, in this case the strikes by an AC-130 and attack helicopters had precisely the opposite result: coercive strikes caused coalition fragmentation, undermining the threat of damage yet to come.

COALITIONS AND ADVERSARY COUNTER-COERCION

Not surprisingly, foes try to shatter coalitions. Because the United States exhibits a strong desire to conduct coercive operations as part of a coalition rather than unilaterally, coalition unity itself becomes a vulnerable center of gravity that adversaries attempt to exploit. Exploiting coalition fissures offers adversaries an enticing counter-coercive strategy, as an alternative or adjunct to combating threats of force directly. Saddam Hussein attempted to widen coalition splits at several key junctures in the Gulf crisis and its aftermath, in an effort to undermine the threat of escalation against Iraq. Prior to the coalition ground assault, his attempted negotiations with the Soviet Union had the effect of not only nearly averting war but also causing some coalition members to question the need for military action. "[O]nce forged, the U.S. unwritten alliance with the Soviet Union itself became a center of gravity,"[28] because unrelenting U.S. efforts to build a broad East-West and North-South coalition signaled Washington's desire to avoid destroying its bilateral relations with key states. Iraq similarly tried to dislodge Arab support for coalition operations by linking resolution of the Kuwaiti crisis to the Arab-Israeli dispute, thereby driving a wedge between the Arab states and the U.S.-Israeli axis.

---

[28]Summers (1992) p. 233.

Adversary efforts to undermine coalition unity can disrupt coercive operations in several ways. First, to the extent that the coalition itself amplifies the coercive threat by aggregating military assets, rupturing coalition relations can cause the withdrawal of members' contributions, which may be vital when access and basing are limited. Second, internal coalition disagreements can force those members most willing to escalate to lower their own ceilings of practicable force to repair coalition unity. At the January 1994 NATO summit, where disagreements surfaced between the United States and its European allies over the use of air strikes, diplomats forged a compromise declaration that appeared to limit when such strikes would be contemplated, trading off a reduced coercive threat for coalition cohesion.

In addition to causing a coalition to limit the amount of force it is willing to employ, adversary efforts to split it can shape the conduct of coercive operations when force is applied. As noted above, Saddam viewed the fragility of Arab-Israeli relations as potentially exploitable. Iraqi Scud attacks against Israel may have reflected calculations that drawing Israel into the conflict would destroy coalition unity by driving Arab states to withdraw. In response, coalition planners diverted American air power from other operations to seek and destroy Scud launchers, hoping to stave off Israeli intervention.[29] Here, the flexibility of air power proved useful for bolstering coalition support—the ability to reorient air power to different missions quickly can help avert internal coalition disputes, although the tradeoff may be a reduction in overall coercive potency.

The potentially disruptive effects of coalition fissures on coercion means that even relatively minor actions by the adversary can have enormous strategic implications. Low-level violence by Aideed's

---

[29]At the tactical level, too, coalition air forces brought their own sets of preferences. As the *Gulf War Air Power Survey* reported:

> Allied cooperation did not . . . simply make Coalition air forces extensions of the U.S. Air Force. The governments concerned kept control over the targets that their forces could strike; on the whole, the limitations were neither burdensome to the Joint Force Air Component Commander (JFACC) nor substantially different from those imposed on American forces. Nonetheless, the weapon systems foreign air forces favored did at times shape Coalition tactics. (Keaney and Cohen, 1993, p. 159.)

faction following air assaults on his compounds caused a major crisis within the coalition. Iraqi Scud attacks against Israel were, from a purely military standpoint, of minimal effectiveness. Had they succeeded in drawing Israel into the war or induced Israel to launch its own defensive air strikes, Arab support for the war might have collapsed, threatening overall U.S. regional strategy.

CONCLUSIONS

The effects of coalition-building on coercion confront policymakers with a balancing calculation. The United States may, for a variety of reasons, need to build and maintain coalitions in conducting military operations, but at some point the marginal benefit of further coalition-building—whether in terms of size or degree of cooperation and unity—may level off and even turn negative. This reality has significant implications for military planners, because coercive U.S. air power will often be used in coalition contexts, and also because air power may offer opportunities to mitigate tensions between coalition maintenance and coercion.

Coercion is not simply a product of relations between a coercer and an adversary but also of the dynamics within the coercer. Chapters Four and Five illustrate how factors internal to the coercer—domestic public opinion and coalition unity—shape coercive threats and an adversary's perception of them. As the following chapter shows, dynamics internal to the adversary have similarly strong effects.

Chapter Six

COERCING NONSTATE ACTORS: A CHALLENGE
FOR THE FUTURE

Humanitarian operations and crises involving confrontations with nonstate actors—communal militias, violent political movements, and other organized political actors that are not nation-states—are increasingly common in the post–Cold War world. In 1991, the United States intervened in post–Operation Desert Storm fighting in Iraq, providing aid and assistance to the country's Kurdish and Shi'a populations. In 1992 and 1993, it sent combat troops to help stave off a humanitarian disaster in Somalia. And in 1995, it deployed forces to the former Yugoslavia to solidify a peace agreement between rival ethnic groups.

Coercion will be a critical foreign policy tool in crises involving nonstate actors. The United States will turn to military force because many nonmilitary forms of pressure, such as economic sanctions and diplomatic efforts, are difficult to target against nonstate adversaries. At the same time, crises will often involve issues that do not directly implicate vital U.S. interests; more frequently, they will involve interests perceived as peripheral to the American public, and will therefore demand strictly limited, as opposed to overwhelming and brute, uses of force.

This chapter describes two common missions involving nonstate actors: coercing the nonstate actor directly and coercing its state sponsor. It then describes several common characteristics of nonstate actors that make them more difficult to coerce. The evidence suggests that difficulties encountered are rarely unique to nonstate actors; they are often present when coercing state actors as well. These problems, however, are often exacerbated in the nonstate context, and therefore deserve separate analysis and elaboration.   It is difficult to generalize about nonstate threats.   Although adversary states differ in a number of important attributes, the spectrum of potential nonstate actor threats is virtually limitless.   This chapter therefore examines a variety of recent crises involving attempts to coerce nonstate adversaries to illustrate a wide range of issues associated with such strategies.

TYPES OF MISSIONS

Conflicts with nonstate actors involve a wide range of interests and military missions, from humanitarian operations to those related to guerrilla and terrorist groups.   But the coercion of nonstate actors typically involves coercing local warlords and the sponsors of nonstate actors to accede to a variety of demands.

Coercing Local Warlords

The United States has been called on to coerce local warlords who have threatened the security of U.S. and allied citizens or the citizens of their own country.   Such a task is particularly common during humanitarian operations such as in Bosnia, Somalia, and Rwanda, where local warlords interfered with the success of the mission.

Aiding humanitarian relief efforts is a common mission that itself may not require coercion.   In Bangladesh, the United States provided vital relief following Cyclone "Marian" in 1991—in Operation Sea Angel—without incurring opposition.   Many humanitarian operations, however, are not so straightforward.   Nonstate actors sometimes interfere with the distribution of humanitarian relief, requiring the intervening power to intimidate them into cooperation or at least noninterference.   As central authority broke down and civil war spread in Somalia, the resulting anarchy allowed widespread banditry and looting of relief supplies; it also presented rival clan leaderships the opportunity to exploit control over vital delivery routes and to extort profits to enhance their power bases.   In spring 1992, the UN authorized a relief mission, the UN Operation in Somalia (UNOSOM I).   The operation included a small peacekeeping force to establish a secure environment for humanitarian relief organizations to carry out their functions. The UN presence found itself ill-equipped to secure transportation of aid through the many armed, tribal bands which themselves lacked centralized control. The humanitarian mission was failing and the warlords would not cooperate.

In late 1992, the UN Security Council responded to the failure of UNOSOM I by authorizing a more militarily robust intervention by a United States-led coalition, the Unified Task Force (UNITAF). The initial phase of Operation Restore Hope included 28,000 U.S. servicemen and considerable combat potential.[1] In a more aggressive approach, UNITAF began limited efforts to disarm the various factions that posed threats to humanitarian aid. UNITAF planners concluded that force, or a credible threat of force, was required to ensure the safe distribution of food, particularly in the "triangle of death" south-central region, which was largely under the control of Somali warlord Mohammed Farah Aideed.[2] Because the violence endemic to the region was carried out by individuals and local militias aligned largely according to clan loyalty, and because the Somali lifestyle valued self-reliance highly, UNITAF disarmament efforts aimed both carrots and sticks at the individual Somali: UNITAF initiated small-scale, weapon-exchange incentive programs as well as more comprehensive confiscation policies (particularly directed at crew-served weapons and heavily armed vehicles).[3]

In summer 1993, UNITAF handed responsibility over to a second UN force, UNOSOM II. Relations between UNOSOM II and Aideed quickly broke down. UN planners likely miscalculated the extent to which Aideed would perceive peacekeeping operations as a threat to his emergent authority within interclan political rivalries. Whereas the UN traditionally engaged in peacekeeping efforts at the invitation of host governments, there was no host government with which the UN could officially negotiate consensual terms. Envoys therefore had to manage precarious relations with various rival factions.

---

[1]U.S. deployments included a Marine Expeditionary Force, the 10th Mountain Division, Air Force and Navy units, and special operations forces including psychological operations and civil affairs units.

[2]Clarke (1993–1994), pp. 45–46.

[3]Lorenz (1993–1994), pp. 30–32.

Aideed perceived UN actions—especially attempts to seize his heavy weapons—as intended to marginalize him, and UN efforts eroded his power base, which had relied heavily on profits gained from looting humanitarian aid.[4] As discussed in the previous chapter, the result was an escalating spiral of violence between UN and Aideed's forces as the UN launched military operations to detain Aideed or to coerce him to comply with UN efforts.[5]

The Somalia experience illustrates that warlords often continue fighting amid a humanitarian disaster and see the aid as a threat. Outsiders thus face the twin challenges of stopping intrastate fighting and providing aid. Each task, in turn, poses complications for coercion; these difficulties are compounded when the tasks are concurrent.[6]

Coercing State Sponsors

Although nonstate actors may themselves be aggressors or otherwise pose a danger to U.S. and allied interests, these actors often receive state backing. Outside powers regularly meddle in civil wars, supporting irredentist or secessionist movements or simply trying to offset the meddling of other powers. In addition, outside governments sponsor communal militias to advance their foreign policies. Rather than threatening a nonstate actor directly, a coercer can threaten its state patron, thereby reducing outside support or leading the sponsor to crack down on the nonstate actor's activities. In essence, this

---

[4]Absent a centralized state, several prominent warlords governed various regions of Somalia with shifting boundaries. Even "neutral" international intervention would inevitably affect this balance of power. Since Aideed saw himself as poised to overturn the status quo balance and assert greater personal authority, he naturally perceived UN stabilization efforts as an obstacle to his objectives.

[5]On June 5, Aideed's forces ambushed UNOSOM II peacekeepers, resulting in the death of 23 Pakistani soldiers. The UN responded by calling for the arrest of Aideed and his allies. U.S. forces and Cobra gunships conducted several military strikes against Aideed's Somali National Alliance (SNA) strongholds, further provoking anti-UN hostility among the Somali people. The strikes included a June 17 attack with AC-130 Spectre gunships on Aideed's residence/command bunker and a July 12 attack by U.S. Cobra gunships on the house owned by Aideed's defense minister, where intelligence sources reported top Aideed aides were meeting. Lippman and Gellman (1993), p. A1; Richburg (1993b), p. A1; and Tubbs (1997), p. 33.

[6]Pirnie and Simons (1996a), p. 16.

is a second-order coercive strategy that requires coercing the sponsor to coerce the nonstate actor—an inherently difficult undertaking.

The conflict in the former Yugoslavia illustrates the relationship between nonstate actors and state sponsors. President Slobodan Milosevic, operating out of the Serbian capital of Belgrade, was the original architect and primary manager of the Serb war effort, even though his authority was based on loosely established lines of command and loyalties in place before the dissolution of Yugoslavia. In May 1992, the Yugoslav People's Army (YPA) split into the Army of Yugoslavia (YA) and the Serbian Army in Bosnia, which later became the Bosnian Serb Army (BSA). The BSA continued to consult closely with its parent organization throughout the conflict, and it received supplies and operational support.[7] Belgrade's influence over military operations in Bosnia derived from this military and other forms of support. As a result, the United States and its allies employed a variety of means to pressure Belgrade in the hope that Milosevic would, in turn, squeeze the BSA.

Perhaps the best illustration of the successful coercion of a state sponsor is Israel's attacks on Palestinians in Jordan during the 1950s. Israel recognized that the terrorism itself could not be stopped by Israeli actions, and that a third-party host was better positioned to control activities from within its territory. As Moshe Dayan declared about Israel's policy in the early days of the state's existence:

> We cannot guard every water pipeline from explosion and every tree from uprooting. We cannot prevent every murder of a worker in an orchard or a family in their beds. But it is in our power to set a high price on our blood, a price too high for the Arab community, the Arab army, or the Arab government to think it worth paying. We can see to it that the Arab villages oppose the raiding bands that pass through them, rather than give them assistance. It is in our power to see that Arab military commanders prefer a strict performance of their obligation to police the frontiers rather than suffer defeat in clashes with our units.[8]

---

[7]Gow (1993), pp. 243–246; Vego (1992), pp. 445–446.

[8]Dayan (1968) as quoted in Bar-Joseph (1998), p. 152.

Israel relied on third parties—Arab military commanders—to restrain movements that Israel itself could not stop. Israeli reprisals in the 1950s succeeded—after several years of unsuccessful attempts to stop infiltration that led to 100 casualties a year from 1951 to 1954—in forcing the Jordanian government to stop Palestinian infiltration. Israeli reprisals against refugee camps and villages in Jordan led to demonstrations against the Jordanian government for failing to protect them.[9] Although King Hussein became militantly anti-Israel in his public diplomacy, at the same time he ordered the army to crack down on any infiltration to prevent domestic unrest. After 1954, infiltration fell dramatically. Israeli raids had threatened King Hussein's quest for national integration, prompting him to seek the *status quo ante.*[10]

Jordan became a key base of Palestinian operations again after the 1967 war. To stop the attacks, Tel Aviv once more relied on a combination of direct strikes on Palestinian targets and pressing the Jordanian government. As in the 1950s, this back and forth created the specter of instability in Jordan. If the Palestinians had been allowed to expand recruitment in Jordan and defend themselves vigorously, they might have become stronger than the Jordanian government itself. The result was "Black September," the month King Hussein cracked down on radical Palestinian activity in 1970 and drove the Palestinian movement outside his borders to Lebanon.

The option of coercing state sponsors will likely be constrained, however, in many cases. Some nonstate threats will not draw substantial support from states (for example, Aideed). Targeting state sponsors may be politically or diplomatically impossible. And many nonstate actors have multiple or ambiguous state sponsors (in the 1970s, the Palestine Liberation Organization [PLO] drew financial support from a large number of Arab states as well as many private citizens). As a result, the United States may choose to coerce the nonstate actor directly.

---

[9]Israel struck primarily at Arab military objectives instead of towns and villages after attacks on Palestinian civilians in Jordan led to condemnation in Israel, the United States, and elsewhere. Morris (1997), pp. 274–276.

[10]Shimshoni (1988), pp. 37–51; Morris (1997), pp. 100–101.

CHARACTERISTICS OF COERCIVE OPERATIONS AGAINST NONSTATE ACTORS

As the above mission descriptions highlight, coercing nonstate actors is both important and complex. Drawing on the lessons identified earlier and applying them to the context of nonstate adversaries, additional insights emerge. Several key features of nonstate actors affect the conditions and challenges identified in Chapter Three and, ultimately, the success of coercion.

Characteristics that distinguish attempts to coerce nonstate actors include:

- Nonstate adversaries may lack identifiable and targetable assets.

- Inaccurate intelligence estimates are particularly common.

- Nonstate adversaries may lack control over constituent elements.

- Indirect coercion is often difficult, unreliable, and counter-productive .

- Nonstate actors are adept at exploiting countermeasures to coercion.

 Most of these problems are not unique to nonstate actors, but they have shown themselves to be magnified in the nonstate context.

Nonstate Adversaries May Lack Identifiable and Targetable Assets

Coercion assumes an ability to hold some adversary interest at risk. For a variety of reasons, the nonstate context complicates this core assumption. Military forces and territory are less often vulnerabilities of nonstate actors. The August 1998 missile attacks against terrorist financier Usama bin Laden illustrate this problem. The target was bin Laden's "network," but it was not clear what this comprised beyond the people involved, because he had few assets associated with the network that were vulnerable to military force.[11] Similarly, the Chechens presented no major military targets for the Russian air force.[12] Compared with many nonstate actors, the Bosnian Serb military was relatively sophisticated. Operation Deliberate Force planners targeted infrastructure and communications networks seen as critical to Serb military effectiveness.[13] Against a less military-technologically sophisticated adversary, such target sets will not be available. In addition, nonstate actors may be less susceptible to coercive threats to armed forces or territory than state actors if their power and legitimacy do not rest on control over that territory.[14] Threats to an adversary's territory, population, and economic well-being are sometiimes elements of coercion, but these can mean little to guerrilla groups. Defeating an adversary's military strategy (denial) is far easier when that strategy is conventional—insurgency operations are by nature less resource-intensive than conventional ones and neutralizing them requires far more time. Even after a devastating military defeat, a nonstate actor can survive as a political institution and revive its armed forces for a continued guerrilla war.

The case of Somalia most clearly illustrates that nonstate adversaries may not possess the multitude of targetable assets possessed by state actors. Aideed's military assets consisted of little more than several thousand militiamen and a few hundred "technicals"—or vehicles equipped with machine guns, antiaircraft guns, or recoilless rifles.[15]

---

[11]The missile attacks on bin Laden fall on the "brute force" end of the coercion spectrum, as their ostensible goal was to remove capabilities by killing people rather than to coerce change in behavior.

[12]Lambeth (1996), p. 365.

[13]Atkinson (1995), p. A1; Covault (1995), p. 27. The costs inflicted by these strikes cannot be measured simply by looking at the targeted assets. Their value lay in magnifying the threat to the Bosnian Serbs posed by the simultaneous Croat and Muslim ground offensives.

[14]In Sri Lanka, government forces believed that capturing key territorial strongholds of the rebel Liberation Tigers of Tamil Eelam (LTTE) would force a favorable negotiated settlement; instead, their operationally successful offensive failed to bring such a result, when the movement proved more adaptable and less reliant on territorial control than predicted. Harris (1996), p. 56.

[15]Intelligence estimates put Aideed's forces at about 5000 men, several hundred of which were ardent supporters constituting his key forces. In addition to "technicals," these forces possessed small arms, limited quantities of artillery and old, Soviet-model tanks. Richburg (1993a), p. A14; Perlez (1992), p. 14.

As an undeveloped country, Somalia lacked military or administrative targets valuable to Aideed. UN planners were limited to targeting Aideed himself (along with his closest advisors), his SNA headquarters, and an SNA-operated radio station. Given that Somali society was already in a state of chaos, there was little that could be held at risk by UN military forces.

This lack of targets can limit the utility of air power when trying to coerce nonstate actors. In Rwanda, President Clinton noted that, unlike the former Yugoslavia, air strikes were not feasible: "Here you had neighbors going from house to house cutting people up with machetes. Who was there to bomb?"[16] Russia faced this problem in Chechnya, where the Chechens avoided any direct challenge to Russia's command of the air. After the pitiful Chechen "air force" was destroyed by Russian forces, Chechen leader Dudayev mockingly congratulated the Russian commander, wiring him a message that read "I congratulate you and the Russian [Air Force] on another victory in achieving air superiority over the Chechen Republic. Will see you on the ground."[17]

The operational concepts used to coerce state actors generally assume an ability to discriminate between military and civilian targets, but this is difficult when confronting nonstate actors. As Chapter Four illuminated, perceived public sensitivity, both at home and abroad, to civilian casualties at times requires that coercive operations avoid damage to civilian lives and property. This is particularly true when the mission is humanitarian. Like most of the problems addressed in this section, the issue of target discrimination is not unique to nonstate actors. However, it is likely to be exacerbated in contexts where the adversary lacks a professional military (which may be identifiable by official markings) and particularly where nonstate actors operate within highly militarized societies. In Somalia and southern Lebanon, for example, the UN and Israel respectively faced enemy personnel virtually indistinguishable from the heavily armed civilian populace.[18] As discussed below, this presents non-

---

[16]Shogren (1998).

[17]As quoted in Lambeth (1996), p. 370.

[18]Schow (1995), p. 23.

state actors with potentially effective countermeasures to coercive air strategies.

Inaccurate Intelligence Estimates Are Particularly Common

The intelligence challenges identified in Chapter Three are particularly acute with regard to nonstate actors. Often the groups in question are poorly known to the West before a crisis occurs. Intelligence about Somali warlord Aideed was extremely limited, making it difficult to track his whereabouts.[19] In addition, the nonstate nature can reduce the availability of even basic information: the United States does not have diplomats, businessmen, or cultural figures visiting, and learning about, "Hezbollahland."

Underestimating or misunderstanding nonstate adversary motivations is particularly likely. Even if a nonstate actor is weak, its motivations are likely to be strong, particularly when compared with those of the coercing power. The perceived benefits of resisting coercive threats are likely to be considerable. In civil war or ethnic conflict, the parties will have already resolved to accept extremely high costs in pursuit of their goals. In the case of religious or ideological movements, nonstate organizations may be driven by intense desires to achieve more transcendent objectives. And in humanitarian crises, violence may stem from perceived necessities of survival. In all of these situations, the United States is likely to face adversaries highly motivated to absorb costs. Whereas nonstate crises will often implicate interests seen as peripheral to the United States and its allies, they may implicate the highest stakes for nonstate adversaries.

Indeed, the coercing power's entry into a conflict often changes the political dynamic of an entire country, making resistance more probable. A segment of the population may not welcome an outsider's intervention and instead may laud obstruction of the intervening power. Aideed's violent responses to UN coercive pressure immediately enhanced his stature within Somalia. Similarly, the Israeli attacks on Hezbollah increased the movement's credibility with the anti-Israel, although not pro-Hezbollah, Lebanese populace.

---

[19]Smith (1993), p. A19.

In formulating coercive strategies against nonstate actors, it is also difficult to establish causal links between identifiable assets and an adversary's cost-benefit calculations. The lack of institutionalized and formal state structure may mean that a substate adversary will prove more resilient than expected with respect to seemingly vulnerable assets or nodes. Russian efforts to eliminate Dudayev, the Chechen separatist leader, in 1996 were premised on the belief that the guerrilla organization depended heavily on Dudayev's personal leadership; after his death, the organization survived and adapted.[20] Somali factions were organized around clan loyalties. Even without a charismatic leader like Aideed, it is likely that another figure would have filled the leadership void. In Hezbollah, Israel faced a diffuse target. Its structure was only partially known to Israeli planners, complicating the process of finding, and then threatening, key organizational nodes.

Nonstate Adversaries May Lack Control over Constituent Elements

Nonstate actors are less likely to control their constituents and agents and thus often cannot make or implement concessions. Because they may lack formal or well-institutionalized control and decisionmaking structures, the lines of authority within nonstate actors can blur or break. Altering the adversary leadership's cost-benefit calculus may therefore not generate the desired changes in behavior by subordinate agents. More broadly, even when coercion has its usual desired primary effects—persuading the adversary *leadership* to change course—these effects may not translate into the desired secondary effect—*compliance.*

In war, disrupting or paralyzing an adversary's command and control often contributes directly to success. In coercion, in contrast, disruption or paralysis can impede success by delaying or preventing full compliance. Similar difficulties inhere to coercing nonstate actors when they lack well-entrenched lines of authority. The PLO proved far easier for Israel to coerce than Hezbollah in the early 1980s, in part because the PLO functioned as a state within a state in

---

[20]"Chechnya After Dudayev" (1996), pp. 1–2.

Lebanon. As Ze'ev Schiff and Ehud Ya'ari contend, "Paradoxically, the more the PLO prospered through the 1970s, the more vulnerable it became, if only because it had more to lose than ever before from any threat to its new stability in Lebanon."[21]

The Serbia-Bosnian entity indicates the reverse phenomenon: it operated much like a state at the outset of the Yugoslav conflict but gradually became less centralized. At the beginning of the conflict, the initial allocation of military resources set up a series of dependency relationships among the various levels in the overall organization. As the conflict intensified, however, this hierarchical structure appeared to suffer from disrupted chains of command. Radovan Karadzic, heading the Bosnian Serb political leadership in Pale, and his self-styled government continually strove to circumvent the control of Serbian President Slobodan Milosevic, even though the Bosnian Serb war effort remained largely dependent on its benefactor in Belgrade. Similarly, the Bosnian Serb military leadership, particularly the senior commander, General Ratko Mladic, frequently defied the Bosnian Serb political leadership. Finally, General Mladic often seemed to lack control over individual Bosnian Serb military officers and militia units.

The Serb leadership was therefore able on several occasions to avert the launching of NATO strikes by claiming lack of control over certain military units. Following air strikes or the threat of them, the Serb leadership could comply satisfactorily with NATO and UN demands while some Serb agents remained noncompliant. In a sense, so-called "renegade" units remained insulated from NATO strikes because NATO's coercive strategy aimed almost exclusively at altering the Serb leadership's cost-benefit calculation. At the same time, the "dislocation of authority" insulated those at the top from the threat of follow-on, escalatory strikes—the Serb leadership could comply with Western demands while its agents ignored them.[22]

Operation Deliberate Force further illustrates how the resulting multiheaded structure can degrade the effectiveness of coercive threats. By September 4, 1995, air strikes appear to have had their in-

---

[21]Schiff and Ya'ari (1984), p. 79.

[22]The challenges posed by "dislocation of authority" for coercive strategies are discussed in Waxman (1997b).

tended *direct* effects: the Bosnian Serb political leadership issued a written commitment to pull back heavy weapons from around Sarajevo. For the next several weeks, however, General Mladic refused to withdraw his forces. The siege of Sarajevo continued and the Western powers were forced to escalate the intensity of their air campaign.[23] NATO strikes successfully altered decisionmaking at the political leadership level, but the organizational structure of the adversary impeded transmission and execution of these decisions. Even as costs of maintaining the siege mounted in the eyes of the political leadership, prompting an agreement to comply with Western terms, the effects did not trickle down in the way coercion theory traditionally assumes. Mladic eventually complied, though not before raising the costs to all parties.

The Israeli experience with Hezbollah and the PLO within Lebanon illustrates a related but distinct challenge in confronting nonstate actors: the inability to make concessions without losing power. Recall from Chapter Three that the costs of acquiescence to a coercer's demand can be prohibitively high, especially in noninstitutionalized democracies where a compliant regime may fear for its very survival. In the early 1970s, the PLO had few high-value targets in Lebanon. More important, Israeli military strikes actually helped PLO recruitment by demonstrating its commitment to the struggle against the Zionist Israel. If the PLO refrained from attacks, other Palestinian groups would gain recruits. Both Hezbollah and the PLO faced constant political competition from rivals within their communities. Any leadership concessions to the Israelis were fiercely criticized and often caused a loss of popular support.[24] Thus, the Israelis risked obtaining concessions that would be meaningless when rivals quickly denounced them.

The structure of nonstate organizations may change as a result of coercive strikes, making it harder to coerce them or to secure im-

---

[23]Hedges (1995), p. 10; Pomfret (1996), p. A24.

[24]Defying coercive threats also provides a way for radicals within a nonstate group to show their disapproval of the dominant group. The PLO was often cautious in its dealings with Israel. More radical groups, such as the Popular Front for the Liberation of Palestinian and smaller splinter groups, used their defiance of Israel to embarrass the PLO, hoping to force the PLO's leadership to choose between kowtowing to Israel and the loyalty of their own supporters.

plementation. In extreme cases, tenuous lines of authority may become severed in the face of coercive threats. The impact may even be counterproductive in the long run. The above examples of competing power centers—such as that between the Milosevic and Karadzic governments and between Karadzic and his military leadership (Mladic)—highlight the potentially adverse interactions of coercive threats when directed at a multiheaded adversary. Belgrade's August 1994 decision to end political and economic support to the war effort marked a split between Belgrade and Pale. At that point, Milosevic's future became tied more closely with preserving his domestic power base and placating the international community than with propagating the war in Bosnia. Karadzic's position, by contrast, became linked even more strongly to perceived adherence to the ultranationalist banner.[25] Similarly, as commander of Bosnian Serb forces, Mladic's stature was based largely on the conflict itself. Resolution of the conflict, barring a clear-cut Serb victory, would undermine the very basis of his authority.[26] The divergent interests in turn conditioned the various centers of power to react very differently to coercive threats. Coercion, designed to induce submission on the part of Serbia, might play into the hands of its rival, the Bosnian Serb political leadership, which could exploit Belgrade's capitulation to harness nationalist sympathies among the population. The simplified, though illustrative, circumstances described here also help explain why coercive threats and air strikes can exacerbate the breakdown of chains of authority between components of the adversary's structure. When the heads of a multiheaded structure have divergent interests, coercive threats may pull them further apart.

Indirect Coercion Is Often Difficult, Unreliable ,and Counterproductive

The cases examined in this study suggest that indirect coercion by promoting third-party threats to the nonstate actor—whether from its rivals or from a government—is potentially effective against state

---

[25]Stieger (1994), pp. 23–24; Kinzer (1994); Djilas (1994), p. 11; Silber (1994), p. 2.

[26]"Ratko Refuses to Leave the Sinking Ship" (1995), p. 57.

and nonstate actors alike, but can more easily spin out of control in the nonstate context.

Because of their relative military weakness, many nonstate groups are highly vulnerable to even poorly armed and organized rivals. Operation Deliberate Force demonstrated that nonstate actors, like their state actor counterparts, are susceptible to coercion when coercive threats magnify third-party military threats:

> Militarily, Deliberate Force was an excellent example of using airpower coercively, to get the Serbs to lift the siege of Sarajevo. For the first 48 hours, NATO aircraft bombed key military targets around Pale with an overabundance of force and were generally impervious to Serb retaliation. . . . Hitting communication nodes, weapons and ammunition storage areas, and lines of communication took away Serb mobility and did not allow them to respond to . . . offensives elsewhere in Bosnia.[27]

Similarly, the Jordanian government quashed Palestinian activity in Jordan after Israeli operations mounted. In each of these cases, however, the coercing power needed a sustained effort to succeed. Israel took years to stop Palestinian cross-border activity, and it flared up anew after years of relative passivity. Similarly, in Yugoslavia the Croat and Bosnian armed forces required several years to mobilize, arm, and train.

Promoting a government crackdown can backfire, however, when the government in question is too weak to control the resulting instability. Israel's effort to force the Lebanese government to quash the Palestinians failed because the Lebanese government, in contrast to that of King Hussein in Jordan, could not provide security. Maronite Christian officers led the Lebanese army into clashes with Palestinian commandos but, by 1969, the army was forced to retreat and give the PLO *de facto* military autonomy in the so-called Cairo Agreement. At the same time, a change of government among the Maronite factions in 1970 resulted in purges of the army and intelligence services, reducing information on Palestinian commandos.[28] In fact, Israeli efforts to prompt a crackdown only highlighted the

---

[27]Beale (1997), p. 37.

[28]Hiro (1992), p. 13.

weakness of the Lebanese government, leading other communal groups to take up arms and hastening the onset of civil war. Thus, the "mechanism" intended by Israel succeeded, but the final result failed.

This tactic can backfire even when the state remains intact. Just as coercive threats or strikes can risk buttressing a state adversary's leadership stature at home and abroad, they can inadvertently increase the support a nonstate organization receives from sympathetic international and local sponsors. Israeli strikes helped Hezbollah attract more money from abroad,[29] and provoked a nationalist backlash, strengthening Hezbollah within the Lebanese community. Israeli military activity and withdrawals from parts of Lebanon in response to Hezbollah violence further bolstered the movement's reputation.[30] Somalia also illustrates similar problems associated with strategies designed to provoke unrest against nonstate actors. The twin objectives—to destroy Aideed's ability to lead resistance to UN efforts while pressuring him to desist—reflected misconceptions of the warlord's position within his factional organization. Traditional clan loyalties would likely have maintained the coherence of Aideed's SNA even without his leadership; among a people that had recently waged a protracted struggle to oust a state regime perceived as illegitimate and sympathetic to imperialism, Aideed gained stature merely by resisting the UN presence.

Working with enemies of nonstate actors can leave the coercer far worse off by strengthening the hands of more radical factions within the nonstate actor. Israeli Air force (IAF) strikes in southern Lebanon provoked a nationalist backlash. While Israel sought to work with moderate militia groups against Hezbollah, air strikes helped rally public support for more radical elements.[31]

---

[29]Schow (1995).

[30]Ranstorp (1997), pp. 38–39.

[31]Because the Lebanese state was weak, and because Hezbollah had tremendous resources and sophisticated social and political networks at its disposal, Hezbollah was able to combine resistance to air strikes with provision of aid to the Shi'a public, further enhancing its standing. After Israel's 1993 Operation Accountability, which caused widespread civilian property damage in southern Lebanon, Hezbollah rebuilt and repaired every damaged building within several weeks, before international organizations could respond. Venter (1996), p. 83.

Nonstate Actors Are Adept at Exploiting Countermeasures to Coercion

Even though nonstate actors may lack institutionalized control of national resources and state infrastructure, they still often possess great ability to employ counter-coercive strategies. In some cases, nonstate actors might even possess greater flexibility and capacity to exploit potential countermeasures than would state actors.

Despite lacking monopoly control over state infrastructure, nonstate actors often have tremendous ability to manipulate domestic and international popular opinion. Aideed was able to garner increased public support by depicting UNOSOM II as yet another foreign effort to dominate the Somali people and exploiting civilian casualties resulting from engagements with UN forces.[32] He was able to do this despite the fact that Somalia lacked high-technology communications for disseminating propaganda (several UNOSOM attacks were directed at an Aideed-controlled radio broadcasting station, used to spread propaganda). Hezbollah successfully depicted Israeli operations as oppressive not only to southern Lebanon's own population but to the international community as well, thereby gaining outside support. Hezbollah has its own public relations office and has proven adept at publicizing its successful operations.[33]

In some cases, the lack of state institutions may present nonstate actors with enhanced opportunities to counteract coercive threats. The lack of state institutions, in particular state military forces, may allow nonstate actors to take advantage of restrictive rules of engagement. In Somalia, Aideed and his followers employed "human shields" to prevent UN reprisals.[34] Nonstate actors are often particularly adept at exploiting human shields and blurring combatant-noncombatant distinctions. In Somalia, the various factions had long organized militia forces according to clan loyalty rather

---

[32]Tubbs (1997), p. 35.

[33]Venter (1996), pp. 81–82.

[34]State actors also employ this technique. Saddam Hussein has used his authoritarian state apparatus with great success to put civilians in harm's way when faced with the threat of air strikes. Crossette (1998), p. A6.

than military professionalism. As Colonel F. M. Lorenz (USMC), the senior legal advisor for Operation Restore Hope, explained:

> Somalis are a nomadic people organized into an extensive clan structure that has existed since the middle ages. The tactics used by the opposing factions were not new. . . . Both [Somali factions] used women and children as active participants. Since women and children were willing participants in the conflict, there was no apparent violation of international law.[35]

The tactics proved easily transferable to conflict with the UN, hindering U.S. and UN efforts to distinguish between combatants and noncombatants. Especially when coercive threats are employed concurrently with humanitarian operations, nonstate adversaries can exploit restrictive ROE to escalate successfully. Requirements such as using minimum force and ceasing fire when hostile forces disengage allow the adversary to push the threshold of retaliatory response.[36]

Nonstate actors are also relatively flexible at countering coercive threats by escalating in unpredictable and unconventional ways. Nonstate actors can rarely escalate in kind, but they can still impose the threat of large costs on militarily far superior coercers. Rather than escalating *vertically*, to match the new degree of violence by the militarily dominant side, the weaker, nonstate power is more likely to escalate *horizontally*, by exploiting the dominant side's vulnerabilities.

Serb forces obviously did not possess the military capabilities to retaliate in-kind to NATO air strikes. However, the ability of the Serbs to counter-coerce the Western powers became readily apparent in April 1993, when NATO began enforcing the "no-fly zone." Although no specific threats were offered by the Serbs, UN aid flights were suspended the day before the first NATO air patrols for fear of reprisals.[37] On several occasions, the Serbs responded to NATO air strikes against military installations by detaining lightly armed

---

[35]Lorenz (1993–1994), p. 36.

[36]Berkowitz (1994), pp. 635–646.

[37]Tanner (1993).

peacekeepers on the ground. In all of these cases, the Serbs threatened the weakest points of the overall UN effort—the vulnerability of humanitarian assistance and ground personnel—to up the ante and deter immediate follow-up strikes. Threats to peacekeepers and to aid flights have tremendous political significance, far greater than their direct military significance. Hence, the Serbs, even without matching the Western powers militarily, were able to manipulate the cost-benefit equation of the UN with relative ease.[38]

Nonstate actors may be more willing to escalate coercive contests by engaging in terrorism than would be state adversaries. The PLO used terrorism to offset Israeli attacks, thereby undercutting Israel's drive to gain escalation dominance through superior conventional might. In the 1970s and 1980s, radical Palestinians hijacked planes and assassinated Israelis overseas and in Israel, killing dozens of Israeli civilians. When Russian forces finally seemed to have taken control of Chechnya, Chechen forces engaged in a number of terrorist acts, including hostage taking, far from the breakaway republic. The Serbs, fearing NATO air strikes, took UN soldiers hostage. As with their ability to exploit political constraints facing coercers, nonstate actors are often well positioned to exact costs of unpredictable kinds and levels against much stronger state actors.

CONCLUSIONS

Nonstate adversaries pose additional challenges for coercion both because of the nature of the actors and the missions often conducted simultaneously with coercive operations. Despite an extremely favorable balance of conventional military power, the United States is likely to face huge obstacles in securing escalation dominance over or denying the strategic objectives of these adversaries. Such actors

---

[38]Of greater import may be the issue that the use of coercive force may undermine the international community's ability to fulfill humanitarian objectives. Humanitarian operations already face great difficulties in war-torn environments because humanitarian aid inevitably benefits certain parties to the conflict. This is especially true in cases such as the former Yugoslavia, where Serbs gained control of territory through denial of sustenance as a means of forcing ethnic civilian population movements. Woodward (1995), p. 319. The difficulties facing humanitarian operations in maintaining an image of impartiality are complicated one step further once coercive force is employed—in Bosnia and Somalia, warring parties did not make a distinction between the UN's humanitarian and military missions.

provide few easy targets to destroy or hold at risk; they can flexibly adapt to or counter military strikes. Working with local, opposing parties (state or nonstate) will often be necessary. These strategies, however, have their own drawbacks and require a more sophisticated understanding of local dynamics and an adversary's internal workings than may be available.

Coercion, and coercive air power more specifically, has proven effective against a number of nonstate adversaries. Yet air power often cannot overcome inherent problems of dislocated authority or a lack of targets to strike. Success in these cases will often require a convergence of factors, many of them far beyond the control of air planners.

PART 4. COERCION AND THE U.S. AIR FORCE

The previous parts of this study suggest that air power can perform a wide range of functions that contribute to coercive diplomacy. Part One put forth a framework for thinking about coercion in the post–Cold War era. Part Two revealed that air power has played an important, and at times decisive, role in many past coercive operations, both by contributing to the conditions that make success more likely and by helping overcome several recurring challenges. Part Three analyzed the context in which coercive air power will be used in the coming years, noting its potential to help smooth differences among coalition members and counter nonstate threats while maintaining the support of the U.S. public. The fourth and final part of this study describes the range of contributions that air power can make to coercive diplomacy and suggests principles to guide its effective future use.

Chapter Seven

IMPLICATIONS AND RECOMMENDATIONS
FOR THE USAF

Air power can play a vital role in successful coercion. Air power's ability to destroy a range of targets, and its growing capabilities in intelligence and precision strike offer new options to military and political decisionmaking. These capabilities, however, do not always lead to more favorable outcomes for the United States. Even if a particular target is destroyed successfully, the change in behavior sought—the true object of coercion—often fails to occur. Understanding this relationship between a target's destruction and the desired outcome is difficult and requires insights into culture, psychology, and organizational behavior.

Air power's unique attributes allow it to play a major role in the three salient factors contributing to successful coercion identified in Part Two: achieving escalation dominance, defeating an adversary's military strategy, and magnifying third-party threats. After addressing each of these in turn, this chapter outlines key ways in which air power can help alleviate common obstacles to successful coercion. The chapter concludes by describing the ways in which air power is especially suited for coercive diplomacy in the present and future security environment and political context. It cautions, however, that the misuse of coercive air power under inauspicious conditions can undermine its future potency and credibility.

AIR POWER AND ESCALATION DOMINANCE

In terms of raw capabilities, air power affords U.S. decisionmakers a range of powerful escalatory options. The Gulf War revealed the awesome potential for modern U.S. air power to destroy a vast array of targets with speed and precision and at low cost in American lives. Air power can help deploy and support ground forces as well as sustain high-intensity combat operations against military targets. Air power can also damage a nation's economic infrastructure, communications network, or other important targets.

Some theorists argue that air power can disrupt the enemy "system," preventing it from executing decisions while inflicting only limited destruction.[1] The United States must recognize, however, that targeting an adversary "system" often fails because adversaries are skilled at adapting their behavior and may not be dependent on a modern industrial infrastructure. Despite U.S. disruption of Iraqi command and control assets, Saddam Hussein remained in touch with his fielded forces through couriers, and Iraq sustained massive economic damage while continuing to defy the United States. This limit to infrastructure and communications strikes applies even more to less-developed nations or nonstate actors. U.S. efforts to destroy North Vietnam's fledgling industrial sites largely succeeded, but they had little true effect because the country's economy relied on agriculture and aid from fellow Communist states. Hezbollah, for its part, does not possess an economic infrastructure in any normal sense of the word. Thus, the United States must recognize that trying to disrupt an adversary's "system" often has a limited effect and that the vulnerability of an adversary to such attacks varies considerably depending on its economic development and political system.

In addition to inflicting damage on an adversary, the USAF can prevent an adversary from escalating and inflicting costs on the United States or its allies. As explained in Part Two, however, escalation dominance is a relative, not an absolute, capability. It implies an ability to impose greater and greater costs on the adversary while denying it the ability to counterescalate. In planning coercive operations, air planners must therefore focus not only on air power's de-

---

[1]This view of the adversary as a "system" has been most famously advocated by John Warden. See Warden (1997/1998) and Warden (1992) for more on this concept.

structive potential, but on its ability to restrict an adversary's escalatory or counter-coercive options. Among the many such contributions that air power can make are suppressing an adversary's ballistic missile or weapons of mass destruction capabilities, providing close air support for peacekeeping or humanitarian personnel who may be vulnerable to reprisals, and protecting the security of allies. Because of its strong air power assets, the United States may prevent an adversary from considering certain options—such as large-scale armored offensives—because it knows the United States could easily defeat these options. In short, air power's greatest impact on successful coercion may not even be observed.

Because adversaries anticipate a coercive response, air strikes, like any coercive instrument, can be ineffective or even counterproductive. *Air power, moreover, is often ineffective precisely because adversaries recognize air power as a U.S. strength and thus plan their provocations accordingly.* When adversaries defy the United States, they adapt their tactics and strategy to fit. Just as skilled military commanders play their strengths against their foes' weaknesses, so too do political leaders defy, provoke, and bargain with the coercer's weaknesses. This conclusion can be labeled "the coercive paradox": the more formidable air power or any other instrument of coercion, the more likely adversaries are to be prepared for it. Adversaries will prepare for it operationally and will avoid strategies that are susceptible to denial by air power.

As Part 3 of this study illustrates, many of the most restrictive constraints on the U.S. ability to use air power to escalate are not technical or operational—they are political and diplomatic. It is these constraints, rather than the technical capabilities of air power, that adversaries are likely to exploit. Domestic public opinion and coalition dynamics frequently place ceilings on the level of practicable force (at least in the eyes of the adversary). Air power can contribute to escalation dominance by helping to overcome some of these constraints on options.

Coalition members have divergent preferences with respect to the use of force. It may be that using U.S. military assets to reduce allies' vulnerabilities, even if those assets are diverted from other functions, is critical to the success of coercion. During Operation Desert Storm, the United States had to devote hundreds of sorties to the "Scud hunt" to reassure Israel—and to prevent Iraq from splitting the coalition arrayed against it—even though these sorties contributed relatively little to defeating the Iraqi military.

Air power can also reduce the restraints imposed by coalitions. Although other major powers continue to modernize their air forces, U.S. capabilities will remain unmatched for years to come. Thus, the primary military role of coalitions involves basing and access, and coalitions will play a political role in enhancing domestic support. Air power places relatively few coalition assets at risk and can help protect against adversary escalation. Because allies often share our sensitivity to friendly casualties, air power can help maintain coalition unity. Steps to reduce U.S. dependence on foreign bases, however, lessen the constraining influence of coalition partners' concerns. Longer-range systems and expeditionary forces also are useful in this regard. Nevertheless, some constraints will endure because of the *political* need for coalitions, regardless of U.S. capabilities for unilateral actions.

Most U.S. military operations for the foreseeable future will be undertaken with limited or less-than-majority American public support. Technological advances that expand the USAF's effectiveness will help it play an important role overcoming possible domestic constraints on the use of force such as casualty sensitivity. The availability of escalatory options that pose little risk of friendly casualties, even if less physically potent than ground or other military options, can help abate public sensitivities and therefore make additional forceful options more credible (although policymakers, in selecting air power as a coercive instrument, must be careful not to signal inadvertently a lack of resolve).

Possible examples of technological advances that might provide the USAF with capabilities that will help overcome or alleviate U.S. domestic constraints include:

- Highly effective unmanned weapons, such as cheap standoff munitions and space-based assets, that pose no risk of U.S. casualties.

- Unmanned aerial vehicles (UAVs) and other forms of surveillance that can enhance U.S. intelligence capabilities with low risk to U.S. lives.

- Tailored munitions that minimize adversary casualties, including nonlethal weapons and munitions that have a focused effect, such as very small laser-guided bombs.

- Systems that provide the ability to insert or retrieve U.S. or allied personnel by air anywhere. They may involve blinding adversaries temporarily or otherwise inhibiting their ability to attack U.S. forces.

- Information warfare assets that allow the United States to disrupt adversary communications and otherwise prevent an adversary from offsetting U.S. coercion efforts.

Such technological advances increase U.S. escalation dominance not only by giving the United States capabilities to impose new costs on adversaries, but also by preventing foes from effectively counter-coercing the United States. In assessing new technological capabilities, policymakers must bear in mind, however, an insight of Chapter Three: that the probability of successful coercion may be more a product of the adversary regime's characteristics or its particular interests—factors over which policymakers have no control—than of the military instrument used.

AIR POWER AND ADVERSARY MILITARY OPERATIONS

One of air power's most important functions—one increasingly practical given continuing advances in intelligence and precision strike capabilities—is threatening an adversary with defeat or otherwise preventing it from achieving its military objectives. Air power can play a key role against state adversaries seeking to impose their regional ambitions via armed aggression before a *fait accompli* occurs.

Aside from the precision and potency of modern U.S. air power, the flexibility and versatility of the air arm suit it well for denying an adversary the perceived fruits of military operations. Air power can be used to disrupt command and control of adversary forces or otherwise conduct strikes that demonstrate the *potential* to devastate—the essence of coercion. In Operation Deliberate Force, NATO air strikes knocked out Serb command and communications facilities with relatively little risk of allied or Serb civilian casualties. Air power can be extremely effective against fielded forces in certain environments. Operation Desert Storm demonstrated this capability vividly, when U.S. air power destroyed parts of two Iraqi corps before they engaged U.S. ground forces near Khafji. The small Iraqi force that did capture the empty town was then isolated and destroyed by coalition ground and air forces. And air power has proven a powerful interdiction tool. During the Linebacker operations in Vietnam, the North Vietnamese discovered that they could not sustain large-scale conventional operations in the face of the damage done by the U.S. bombing campaign. In sum, air power can prevent or hinder certain types of operations, particularly those involving considerable mechanized forces and large logistics efforts.

Air power, of course, is not omnipotent, and its ability to deny an adversary military victory is often limited:

- Air power is less effective against particular types of targets and in particular environments, although technological advances in surveillance, all-weather operations, and precision strike may make air power more potent against these difficult-to-target foes. Adversaries fighting in mountainous, urban, or jungle terrain can often camouflage their movements, making it harder to strike them. Moreover, light infantry units still are difficult for air power to strike.[2]

- In addition, the nature of some adversaries makes them less vulnerable to air operations designed to thwart their military designs. Nonstate actors in particular provide few military targets to destroy or interdict. Their reliance on low-technology communications inhibits air power's ability to paralyze their operations.

- Finally, some adversary *objectives* do not lend themselves to coercive strategies aimed at thwarting military campaigns. Almost by definition, such strategies assume that the adversary is pursuing hegemonic or territorial designs via ongoing armed operations. But the USAF will be called on to compel quite different types of behavior: complying with international mandates, refraining from supporting terrorism, dismantling

---

[2]For ways to improve this capability, see Vick et al. (1996).

WMD programs, or abandoning other goals that do not require the adversary to employ conventional forces.

In sum, air power is an ideal instrument for harnessing the coercive potential of denying many adversaries military victory. However, in many circumstances, some of them growing more likely in the current security environment, air power's role in denial is necessarily limited. Again, adversary strategies are likely to adapt to the strength of air power. As adversaries adapt their strategies to evade denial by air power, the visible role of air power is limited by its own potency.

AIR POWER AND THE MAGNIFICATION OF THIRD-PARTY THREATS

Some of the most notable coercion successes have relied on third-party threats, and their augmentation by air power, to inject critical costs into the adversary's decisionmaking calculus. Magnifying third-party threats has become an even more inviting approach as perceived U.S. casualty intolerance has prompted planners and policymakers to seek strategies that avoid putting American personnel directly in harm's way.

Air power can be particularly effective in shifting the local balance of forces, leaving an adversary vulnerable to a domestic insurgency or to another external adversary. The establishment and maintenance of "no-fly zones" can deprive one side of command of the air, oftentimes removing a critical element of its military prowess. "No-drive zones" enforced by air power greatly reduce an adversary's ability to conduct offensive operations or build up forces. More directly, by interdicting the flow of men and arms to the front, air power can enhance rivals' offensive power. Beyond this combat potential, air power can magnify third-party threats by providing valuable intelligence or supplies.

Working deliberately with third parties carries with it several dangers. At times, these parties may be no more desirable as partners than the ostensible target of coercion; as a result, the true U.S. goal is often to create a balance of power, not to help the third party win. Even more ominously, by making a third-party threat more credible, the United States may inadvertently widen a conflict or foster ethnic cleansing, as adversaries seek to devastate their rivals.

In many ways, promoting internal instability by working with local opposition forces is similar to working with external forces, but the dangers and challenges are often more severe. Local insurgents usually have no training and poor morale, making all but the most basic military operations difficult. The potential drawbacks of ethnic cleansing, widening the conflict, and inciting nationalism are even more likely to hold. At times, an insurgency must be created from whole cloth. Resentment of a regime among an adversary's population may be high, but authoritarian and brutal adversaries—the types of foes that the United States will likely face in the coming decades— are highly skilled at suppressing unrest. Thus, the United States must often stir up resentment, provide a haven for those in opposition and their families, and otherwise nurture a strong opposition. Supply, airlift, and intelligence all will be necessary to sustain operations. Without such support, any insurgents will simply be crushed by the better-armed, better-organized regime forces.[3]

Weakening an adversary vis-à-vis hostile neighbors is also potentially effective, but the necessary circumstances for employing this strategy limit its use. It requires, among other things, that these hostile neighbors exist, and that it is strategically preferable and politically palatable that we assist them.

The operational challenges for the USAF in any effort to magnify third-party threats are considerable. Third-world militaries often lack even basic professionalism. Coordinating intelligence, air strikes, and other combat and command functions will be a constant obstacle, although the speed and flexibility with which air power can be removed or redirected may offer promise for alleviating some of these concerns.

AIR POWER AND COMMON CHALLENGES IN COERCIVE OPERATIONS

Air power makes several clear-cut contributions to intelligence gathering, a key to any successful military operation. The technological sophistication of U.S. airborne and space-based intelligence gather-

---

[3]For the requirements of supporting an insurgency in Iraq, see Byman and Pollack (1998).

ing allows for unparalleled access to information on adversaries' order of battle and deployments. It also allows for better identification and targeting of key adversary military assets.

But advances in reconnaissance and sensing technology should not obscure coercion's other intelligence needs. Planning for coercion requires understanding the means-end chain by which the adversary will concede. When planning a coercive operation, a particularly important task is "target-to-outcome" analysis, recognizing how striking a particular target might affect the desired outcome and then tracing this through to ensure the desired outcome occurs. In practice, this requires a rigorous assessment of each stage of the coercive campaign. Airmen recognize the significance of finding and destroying the target. Less attention is paid, however, to the ultimate effects of destroying a target. Perhaps more important, airmen often focus exclusively on the physical impact of a target's destruction rather than on its psychological effect. As the above discussion suggests, destroying targets at times has little effect on the success of coercion and may even prove counterproductive.[4]

As stated in Chapter Two, analysts often mistakenly view coercion in temporally narrow terms—that coercion occurs at specific instants or moments. Coercion is an ongoing process that unfolds over time. While threatened costs may peak and the players change course suddenly, the beginning and end points of coercive contests are often undefined. Taking a long-term view, there are many phases of coercive contests where judicious use of air power can significantly improve the chances of success. This chapter has outlined the roles that air power can play prior to issuing threats and during threat execution. Air power can also play vital roles during the subsequent implementation phase. Coercive contests do not end when the adversary leadership concedes—"success" requires that the adversary carry out concessions and that it not violate any agreement. For certain sets of coercer demands, air power is ideally suited for observing adversary compliance and monitoring it over the long term.

---

[4]Karl Mueller argues that many leading theorists of air power from Douhet on do not focus on the coercion mechanism. Even the more recent *10 Propositions Regarding Air Power* by Phillip Meilinger notes that "In essence, Air power is Targeting." (Meilinger, 1995, p. 20, as quoted in Mueller, 1998, p. 187.)

Its speed and flexibility mean further that it can provide an enduring specter of follow-up strikes to deter misconduct.

THE NEED FOR RESTRAINT

Air power, like other military instruments, offers little help to or may even hinder coercion under certain circumstances. Coalition partners and the American public frequently view air power as holding the virtue of offering at least the possibility of accomplishing policy objectives while minimizing risks and costs. This view poses a critical challenge for air power. Because the public and allies often see air strikes as a low-risk, low-commitment measure, air power will be called on when U.S. public or allied commitment is weak—a situation that will make coercion far harder. The prospects of escalation will be difficult in such circumstances, ROE are likely to be burdensome, and adversaries will question U.S. credibility. Air power's effectiveness will be perceived to suffer as a result. Again, air power's very strengths with regard to domestic support and coalition dynamics could become weaknesses, if they lead air power to be used in situations that lower its credibility. Such use of air power may damage its credibility in future contexts and make coercion even harder.

Policymakers must recognize when air power is not the appropriate tool for the job. When the only targets available to strike (whether constrained by operational, political, or diplomatic pressures) are of limited value to the adversary, air strikes will do little to coerce. Because such strikes will have little impact, they may reduce the credibility of U.S. threats in both the short and long term.

Coercion rests on the credibility of threats, so coercive strategies must be designed to reinforce perceptions of U.S. willingness to use force. When air power fails to coerce, or when the United States concedes to counterescalation, the damage extends far beyond the immediate crisis. Failure can raise other adversaries' doubts about the sustainability of U.S. coercive pressure. It can also lead allies, or potential allies, to cooperate with, rather than oppose, aggressors. Equally important, the misuse of air power can spawn false conclusions, at home and abroad among potential adversaries and allies, about its true effectiveness as a coercive instrument.

For air power to retain its credibility, and hence its ability to coerce, it must be used with restraint. As Eliot Cohen argues, "American air power has a mystique that it is in the American interest to retain. When presidents use it, they should either hurl it with devastating lethality against a few targets (say, a full-scale meeting of an enemy war cabinet or senior-level military staff) or extensively enough to cause sharp and lasting pain to a military and a society."[5] When air power is used to attempt missions it cannot plausibly fulfill, this perception is diminished and adversaries will be more willing to challenge the United States.

By recognizing when air power is likely to fail and avoiding its use in such circumstances, the USAF will better preserve the credibility of air power for instances when it can coerce successfully.

Air power can deliver potent and credible threats while neutralizing adversary countermoves. When favorable factors such as those identified in Chapter Three are absent, however, air power—or any other military instrument—will probably fail to coerce. Policymakers' use of coercive air power under inauspicious conditions diminishes the changes of using it elsewhere when the prospects of success would be greater.

---

[5]Cohen (1994), p. 124.

CASES EXAMINED IN THIS STUDY

Appendix A

Cases Examined in This Study

Coercer	Adversary	Date	Goal of Coercing Power	Outcome		Key Air Power Roles
				Desired	Undesired	
Britain	Argentina	1982	Regain Falkland Islands	Return of Falklands to Britain	Limited "brute force" required to expel Argentines	Deny enemy air superiority; reconnaissance; defense of fleet
Britain	Malayan guerrillas	1948–1960	Defeat Communist insurgency	Defeat of guerrillas	Protracted effort required	Some bombing and reconnaissance; transport and resupply
Britain, France, and Israel	Egypt	1956	Reverse Suez Canal national-ization; destroy Egyptian military threat; humiliate and topple Nasser	Partial destruction of Egyptian military strength	Nasser's popularity increases; canal remains Egypt's; tension between coercers and United States	Assisted Israeli military campaign and British and French invasion
France	FLN insurgency	1954–1962	Defeat pro-independence movement (FLN) and its armed forces (ALN)	ALN forces devastated	Protracted conflict; large losses; political exhaustion and unrest in France; independent, FLN-dominated Algeria recognized; tension within NATO	Prevent infiltration and rapid lift

Appendix A—continued

Coercer	Adversary	Date	Goal of Coercing Power	Outcome		Key Air Power Roles
				Desired	Undesired	
Iran	Iraq	1974–1975	Gain Iraqi concessions over the border	Iraq concedes to Iranian demands		N.A.
Iraq	Iran	1982–1988	Provoke unrest; reduce support for regime; force Iran to halt ground offensives	Fear of attacks at times causes short-term de-escalation	Little or no unrest; adversary regime popularity may have increased; leads to attacks on Iraqi cities	Attacks on cities, shipping, and refineries
Israel	Egypt	1969–1970	Stop Egyptian harassment along canal	Egypt temporarily moderates territorial goals; Egyptian leaders fear military losses and civilian unrest	Cairo turns to Soviet Union for additional support; Israel begins to lose local air supremacy; disruptions of arms supply from United States	Attacks on military-related industries; counterartillery and suppression of enemy air defense (SEAD)
Israel	Palestinians in Jordan	1950s–1970	Stop Palestinian cross-border attacks and terrorism in general	Palestinian attacks eventually stop after Jordanian government intervention	Protracted effort required; diplomatic difficulties; exodus of the PLO to Lebanon with resultant war and instability	N.A.

Appendix A—continued

Coercer	Adversary	Date	Goal of Coercing Power	Outcome		Key Air Power Roles
				Desired	Undesired	
Israel	Lebanese Hezbollah	1982–present	Stop cross-border attacks and terrorism in general	Hezbollah limits scope and scale of attacks	Attacks continue; Hezbollah grows in stature; attacks contribute to Lebanon's instability; domestic criticism of government in Israel	Strikes on training camps
Israel	Palestinians in Lebanon	1970–1982	Stop cross-border attacks and terrorism	Palestinians limit attacks	Protracted effort required; Palestinians at times gain in stature; contribute to Lebanon's civil war; leads to Israeli invasion and quagmire	Transport, reconnaissance, strikes on training camps; limited strikes on infrastructure targets; leadership strikes
Russia	Chechen guerrillas	1994–1996	Crush independence movement		Heavy military and civilian losses; terrorist attacks in Russia; hardening of support for secession	Bomb cities and rebel bases; attack leaders
United States	Cuba and USSR	1962	Force Soviets to withdraw missiles from Cuba	Soviets withdraw missiles		Intelligence; threaten attacks on missile sites
United States	Dominican Republic	1961–1962	Oust corrupt oligarchy	Government changes		Contribute to show of force

Appendix A—continued

Coercer	Adversary	Date	Goal of Coercing Power	Outcome		Key Air Power Roles
				Desired	Undesired	
United States[a]	Germany	1943–1945	Reduce German desire and ability to continue war	"Brute force" damage aids allied military victory, but morale damage has little impact on surrender decision	Considerable casualties, bomber losses	Attacks on industry and population centers
United States	Haiti junta	1994	Oust Cedras regime in favor of Aristede	Junta leaders step down		Lift; intelligence
United States	Iran	1987–1988	Secure free flow of oil	Iran limits attacks on tankers	Occasional Iranian attacks continue	Intelligence; limited attacks on oil platforms and on Iran's navy
United States	Iraq	1991	Remove Iraq from Kuwait and devastate Iraqi heavy forces	By January, Iraq willing to remove forces	Massive effort required; Baghdad refuses maximal U.S. demands	Lift; air superiority for U.S. forces; intelligence; strikes on fielded forces; strikes on strategic targets
United States	Iraq	1991–1998	Compel Iraqi compliance with UN resolutions	Iraq grudgingly accepts inspections; refrains from regional aggression	Protracted effort required; strains U.S.–regional relations; limited compliance	Enforce no-fly and no-drive zone; provide intelligence on Iraqi deployments; limit Iraqi escalation; strikes to ensure Iraqi compliance

Appendix A—continued

Coercer	Adversary	Date	Goal of Coercing Power	Outcome		Key Air Power Roles
				Desired	Undesired	
United States[b]	North Korea and China	1950–1953	Reduce Communists' desire and ability to continue war	Communists agree to armistice after recognizing it cannot gain military victory	Protracted, bloody effort required; U.S. does not achieve maximal goals	Air interdiction, close air support, and attacks on civilian infrastructure
United States	Laotian guerrillas	1960–1973	Stop North Vietnamese Army from transiting Laos		Protracted effort has limited impact on flow of arms	Intelligence to local Laotian forces; air interdiction
United States	Libya	1986	End Libyan support for terrorism	Enhances credibility with allies regarding counterterrorism	Temporary surge in Libyan-supported terrorist attempts	Bomb regime assets
United States	North Vietnam (Rolling Thunder)	1965–1968	Compel North to stop supporting guerrillas in the South		Protracted effort with little impact; increased Chinese and Soviet aid to North	Interdiction; attack military production targets
United States	North Vietnam (Linebacker I and II)	1972	Bring about cease-fire	Hanoi agrees to a temporary cease-fire	Protracted effort; U.S. scales back goals	Attacks on fielded forces and lines of communications
United States/NATO	Bosian Serb forces (Deny Flight)	1993–1994	Reduce scope and scale of Balkan conflict	Help contain conflict and ensure humanitarian relief	Incomplete compliance; undermined NATO credibility	Enforce "no fly" zone

Appendix A—continued

Coercer	Adversary	Date	Goal of Coercing Power	Outcome		Key Air Power Roles
				Desired	Undesired	
United States/ NATO	Bosnian Serb forces (Deliberate Force)	1995	Compel Serbs to accept cease-fire	Serbs seek negotiated agreement		Attack Serb communications and infrastructure; threaten fielded forces
United States/ UN	Somali factions	1993	Stop interference with humanitarian aid	Secured flow of aid	Increased support for anti-U.S. factions; leads to early U.S. withdrawal	Attack leadership targets

[a]The Allied bomber offensive and operations against Germany are more properly seen as a case of brute force.

[b]Like operations in World War II, operations in Korea fall more on the brute force end of the spectrum than on the coercion end.

CASES AND CONDITIONS FOR SUCCESS

Appendix B
Cases and Conditions for Success

Coercer/ Adversary	Outcome		Escalation Dominance	Threaten Military Strategy	Magnify Other Threats
	Desired	Undesired			
Britain/ Argentina	Return of Falklands to Britain	Limited "brute force" required to expel Argentines	Yes	Yes	
Britain/ Malayan guerrillas	Defeat of guerrillas	Protracted effort required	Yes	Yes	
Britain, Israel, and France/Egypt	Partial reduction of Egyptian military strength	Nasser's popularity increases; canal remains Egypt's; tension between coercers and United States		Yes	Opposite (internal cohesion in Egypt strengthened)
France/FLN	Rebel forces devastated	Protracted conflict; large losses; political exhaustion and unrest in France; independent, FLN-dominated Algeria recognized; tension within NATO		Yes	Yes
Iran/Iraq	Iraq concedes to Iranian demands	Iran attacks Iraqi cities	Yes	Yes	Yes
Iraq/Iran	Fear of attacks at times causes short-term de-escalation	Little or no unrest; adversary regime popularity may have increased			Limited

Appendix B—continued

Coercer/ Adversary	Outcome		Escalation Dominance	Threaten Military Strategy	Magnify Other Threats
	Desired	Undesired			
Israel/Egypt	Egypt temporarily moderates territorial goals; Egyptian leaders fear military losses and civilian unrest	Cairo turns to Soviet Union for additional support; Israel begins to lose local air supremacy; disruption of U.S. arms supply	Opposite (Soviet intervention)	Yes (until Soviet intervention)	Yes
Israel/Jordan	Palestinian attacks eventually stop after Jordanian government intervention	Protracted effort required; diplomatic difficulties; exodus of the PLO to Lebanon with resultant war and instability	Yes		Yes
Israel/ Lebanese Hezbollah	Hezbollah limits scope and scale of attacks	Attacks continue; Hezbollah grows in stature; attacks contribute to Lebanon's instability; increased domestic criticism on Egyptian government			Opposite
Israel/ Palestinians in Lebanon	Palestinians limit attacks	Protracted effort required; Palestinians at times gain in stature; contributes to Lebanon's civil war; leads to Israeli invasion and quagmire			Yes

Appendix B—continued

Coercer/ Adversary	Outcome		Escalation Dominance	Threaten Military Strategy	Magnify Other Threats
	Desired	Undesired			
Russia/ Chechens		Heavy military and civilian losses; terrorist attacks in Russia; hardens support for secession			Opposite
U.S./Cuba and USSR	Soviets withdraw missiles		Yes		
U.S./ Dominican Republic	Government changes		Yes		Yes
U.S./Germany	"Brute force" damage aids allied military victory	"Brute force" damage aids allied military victory, but morale damage has little impact on surrender decision; considerable casualties and aircraft losses			
U.S./Haiti	Junta leaders step down		Yes		
U.S./Iran	Iran limits attacks on tankers	Occasional Iranian attacks continue	Yes	Yes	
U.S./Iraq (1990–1991)	By January, Iraq willing to remove forces	Massive effort required; Baghdad refuses maximal U.S. demands	Yes	Yes	Yes
U.S./Iraq (1991–1998)	Iraq grudgingly accepts inspections; refrains from regional aggression	Protracted effort required; strains U.S.–regional relations; limited compliance	Yes	Yes	Yes

Appendix B—continued

Coercer/ Adversary	Outcome		Escalation Dominance	Threaten Military Strategy	Magnify Other Threats
	Desired	Undesired			
U.S./Korea	Communists agree to armistice after recognizing it cannot gain military victory	Protracted, bloody effort required; U.S. does not achieve maximal goals	Yes	Yes	Yes
U.S./Laotian guerrillas		Protracted effort has limited impact on flow of arms	Yes		
U.S./Libya	Enhances credibility with allies regarding counter-terrorism	Temporary surge in Libyan-supported terrorist attempts	Yes		
U.S./North Vietnam (Rolling Thunder)		Protracted effort with little impact; increased Chinese and Soviet aid to North			
U.S./North Vietnam (Linebacker I and II)	Hanoi agrees to a temporary cease-fire	Protracted effort; U.S. scales back goals		Yes	
U.S./Bosnian Serb forces (Deny Flight)	Helped contain conflict and ensure humanitarian relief	Incomplete; undermined NATO credibility			

Appendix B—continued

Coercer/ Adversary	Outcome		Escalation Dominance	Threaten Military Strategy	Magnify Other Threats
	Desired	Undesired			
U.S./Bosnian Serb forces (Deliberate Force)	Serbs seek negotiated agreement		Yes	Yes	Yes
U.S./Somali guerrillas	Secured flow of aid	Increased support for anti-U.S. factions; leads to early U.S. withdrawal			Opposite

NOTES: Yes = Condition present.

Limited = Condition present to at least some degree.

Opposite = Coercive strikes reduced the condition or created its opposite.

COERCIVE ATTEMPTS AND COMMON CHALLENGES

Appendix C

Coercive Attempts and Coercive Challenges

Coercer/Adversary	Outcome		Intelligence Challenge	Credibility Challenge	Feasibility Challenge
	Desired	Undesired			
Britain/Argentina	Return of Falklands to Britain	Limited "brute force" required to expel Argentines			
Britain/Malayan guerrillas	Defeat of guerrillas	Protracted effort required			
Britain, Israel, and France/Egypt	Partial reduction of Egyptian military strength	Nasser's popularity increases; canal remains Egypt's; tension between coercers and United States	Yes		
France/FLN	Rebel forces devastated	Protracted conflict; large losses; political exhaustion and unrest in France; FLN-dominated Algeria independent, recognized; tensions within NATO		Yes	
Iran/Iraq	Iraq concedes to Iranian demands				

Appendix C—continued

Coercer/Adversary	Outcome		Intelligence Challenge	Credibility Challenge	Feasibility Challenge
	Desired	Undesired			
Iraq/Iran	Fear of attacks at times causes short-term de-escalation	Little or no unrest; adversary regime popularity may have increased; Iran attacks Iraqi cities	Yes		
Israel/Egypt	Egypt temporarily moderates territorial goals; Egyptian leaders fear military losses and civilian unrest	Cairo turns to Soviet Union for additional support; Israel begins to lose local air supremacy; disruption of U.S. arms supply to Israel			
Israel/Jordan	Palestinian attacks eventually stop after Jordanian government intervention	Protracted effort required; diplomatic difficulties; exodus of the PLO to Lebanon with resultant war and instability			Yes
Israel/Lebanese Hezbollah	Hezbollah limits scope and scale of attacks	Attacks continue; Hezbollah grows in stature; attacks contribute to Lebanon's instability; increase in domestic criticism of government	Yes	Yes	

Appendix C—continued

Coercer/ Adversary	Outcome		Intelligence Challenge	Credibility Challenge	Feasibility Challenge
	Desired	Undesired			
Israel/ Palestinians in Lebanon	Palestinians limit attacks	Protracted effort required; Palestinians at times gain in stature; contributes to Lebanon's civil war; leads to Israeli invasion and quagmire		Yes	Yes
Russia/Chechens		Heavy military and civilian losses; terrorist attacks in Russia; hardens support for secession	Yes	Yes	Yes
U.S./Cuba and USSR	Soviets withdraw missiles		Yes		
U.S./ Dominican Republic	Government changes				
U.S./Germany	"Brute force" damage aids allied military victory	"Brute force" damage aids allied military victory, but morale damage has little impact on surrender decision; many casualties and aircraft losses	Yes		Yes
U.S./Haiti	Junta leaders step down				
U.S./Iran	Iran limits attacks on tankers	Occasional Iranian attacks continue		Limited	

Appendix C—continued

Coercer/ Adversary	Outcome		Intelligence Challenge	Credibility Challenge	Feasibility Challenge
	Desired	Undesired			
U.S./Iraq (1990–1991)	By January, Iraq willing to remove forces	Massive effort required; Baghdad refuses maximal U.S. demands	Yes		
U.S./Iraq (1991–1998)	Iraq grudgingly accepts inspections; refrains from regional aggression	Protracted effort required; strains U.S.–regional relations; limited compliance	Yes		Yes
U.S./Korea and China	Communists agree to armistice after recognizing it cannot gain military victory	Protracted, bloody effort required; U.S. does not achieve maximal goals	Yes		
U.S./Laotian guerrillas		Protracted effort has limited impact on flow of arms	Yes		
U.S./Libya ●	Enhances credibility with allies regarding counterterrorism	Temporary surge in Libyan-supported terrorist attempts		Yes	Yes
U.S./North Vietnam (Rolling Thunder)		Protracted effort with little impact; increased Chinese and Soviet aid to North	Yes	Yes	
U.S./North Vietnam (Linebacker I and II)	Hanoi agrees to a temporary cease-fire	Protracted effort; U.S. scales back goals			

Appendix C—continued

| Coercer/Adversary | Outcome | | Intelligence Challenge | Credibility Challenge | Feasibility Challenge |
	Desired	Undesired			
U.S./Bosnian Serb forces (Deny Flight)	Helped contain conflict and ensure humanitarian relief	Incomplete; undermined NATO credibility	Yes		
U.S./Bosnian Serb forces (Deliberate Force)	Serbs seek negotiated agreement			Yes	
U.S./Somali guerrillas	Secured flow of aid	Increased support for anti-U.S. factions; leads to early U.S. withdrawal	Yes	Yes	Yes

NOTES: Yes = Condition present.

Limited = Condition present to at least some degree.

Opposite- Coercive strikes reduced the condition or created its opposite.

BIBLIOGRAPHY

Achen, Christopher H., and Duncan Snidal, "Rational Deterrence Theory and Comparative Case Studies," *World Politics*, Vol. 41, No. 2, January 1989, pp. 143–169.

Air Force Basic Doctrine, AFDD 1, September 1997.

Apple, R. W. Jr., "Clinton Looks Homeward," *New York Times*, January 13, 1994, p. A1.

Arkin, William M., "Baghdad: The Urban Sanctuary in Desert Storm?" *Air Power Journal*, 1997.

Atkinson, Rick, "Air Assaults Set Stage for Broader Role," *Washington Post*, November 15, 1995, p. A1.

Ayoob, Mohammed, "The Security Problematic of the Third World," *World Politics*, Vol. 43, January 1991, pp. 257–283.

Baker, James E., *The Politics of Diplomacy*, G. P. Putnam's Sons, New York, 1995.

Baldwin, David, "The Power of Positive Sanctions," *World Politics*, Vol. 24, No. 1, October 1971, pp. 19–38.

Barabak, Mark Z., "U.S. Raids Get Broad Support: Clinton Issues Not Significant," *Los Angeles Times*, August 23, 1998.

Bar-Joseph, Uri, "Variations on a Theme: The Conceptualization of Deterrence in Israeli Strategic Thinking," *Security Studies*, Vol. 7, No. 3, Spring 1998.

Bar-Siman-Tov, Yaacov, "The War of Attrition, 1969–1970," in Alexander L. George (ed.), *Avoiding War: Problems of Crisis Management*, Westview Press, Boulder, CO, 1991, pp. 320–341.

Beale, Michael O., "Bombs Over Bosnia: The Role of Airpower in Bosnia–Herzegovina," thesis presented to the School of Advanced Airpower Studies, Air University Press, Maxwell Air Force Base, AL, August 1997.

Bennett, D. Scott, and Allan C. Stam III, "The Declining Advantages of Democracy: A Combined Model of War Outcomes and Duration," *Journal of Conflict Resolution*, Vol. 42, No. 3, June 1998, pp. 344–366.

Berkowitz, Bruce, "Rules of Engagement for UN Peacekeeping Forces in Bosnia," *Orbis*, Vol. 38, No. 4, Fall 1994, pp. 635–646.

Bertram, Christoph, "Multilateral Diplomacy and Conflict Resolution," *Survival*, Vol. 37, No. 4, Winter 1995–1996.

Booth, Ken, *Strategy and Ethnocentrism*, Holmes & Meier Publishers, Inc., New York, 1979.

Boudreau, Donald G., "The Bombing of the Osirak Reactor: One Decade Later," *Strategic Analysis*, June 1991.

Brown, John Murray, "Russia, Turkey Voice Concern About New Attacks Against Iraq," *Washington Post*, January 19, 1993, p. A17.

Bueno de Mesquita, Bruce, *The War Trap*, Yale University Press, New Haven, CT, 1981.

Bundy, McGeorge, *Danger and Survival*, Random House, New York, 1988.

Byman, Daniel, and Kenneth Pollack, "Undermine: Supporting the Iraqi Opposition," in Patrick Clawson (ed.), *Iraq Strategy Review: Five Options for U.S. Policy*, Washington Institute for Near East Policy, Washington, DC, 1998.

Byman, Daniel, Kenneth Pollack, and Matthew Waxman, "Coercing Saddam Hussein: Lessons from the Past," *Survival*, Autumn 1998, pp. 127–152.

Byman, Daniel, and Matthew Waxman, "Defeating US Coercion," *Survival*, Vol. 41, No. 2, Summer 1999.

Camerer, Colin, "Individual Decisionmaking," in *The Handbook of Experimental Economics*, Princeton University Press, Princeton, NJ, 1995, pp. 587–703.

"Chechnya After Dudayev," *Strategic Comments*, Vol. 2, No. 4, May 1996, pp. 1–2.

Chubin, Shahram, and Charles Tripp, *Iran and Iraq at War*, Westview Press, Boulder, CO, 1988.

Clarke, Walter S., "Testing the World's Resolve in Somalia," *Parameters*, Vol. 23, No. 4, Winter 1993–1994.

Claude, Inis L. Jr., "The United States and Changing Approaches to National Security and World Order," *Naval War College Review*, Vol. XLVIII, No. 3, Summer 1995.

Clodfelter, Mark, *The Limits of Air Power: The American Bombing of North Vietnam*, Free Press, New York, 1989.

Cohen, Eliot A., "The Mystique of Air Power," *Foreign Affairs*, Vol. 73, No. 1, January/February 1994.

Connelly, Marjorie, "Wide U.S. Support for Air Strikes," *New York Times*, December 18, 1998, p. A26.

Covault, Craig, "NATO Air Strikes Target Serbian Infrastructure," *Aviation Week & Space Technology*, September 11, 1995, p. 27.

Crossette, Barbara, "Civilians Will Be in Harm's Way If Baghdad Is Hit," *New York Times*, January 28, 1998, p. A6.

David, Stephen, *Choosing Sides: Alignment and Realignment in the Third World*, Johns Hopkins University Press, Baltimore, MD, 1991.

Dayan, Moshe, "Why Israel Strikes Back," in Donald Robinson, *Under Fire: Israel's Twenty Year Struggle for Survival*, W. W. Norton, New York, 1968.

Defense News, February 2, 1998, p. 3.

Djilas, Milovan, "Cracks Within the Serbian Monolith," *Guardian*, August 18, 1994, p. 11.

Dupuy, T. N., *Elusive Victory: The Arab Israeli Wars, 1947–1974, 3rd Ed.*, Kendall/Hunt Publishing, Dubuque, IA, 1992.

El Reedy, Abdel Raouf, "Striking the Right Balance," *Guardian Weekly*, January 31, 1993, p. 6.

Eshel, David, "Counterguerrilla Warfare in South Lebanon," *Marine Corps Gazette*, July 1997.

———, "Vintage from 'Grapes of Wrath," *Armed Forces Journal International*, Vol. 133, No. 11, June 1996, p. 22.

Essays on Air and Space Power, Air University Press, Maxwell Air Force Base, AL, 1997.

Fineman, Mark, "Hussein's Moves Seen as Steps in Calculated Plan," *Los Angeles Times*, January 17, 1993, p. A1.

Foreign Relations of the United States 1952–1954. Political Annex prepared by the U.S. Department of State, Policy Planning Staff, June 4, 1953, Vol. XV, Part 1, GPO, Washington, DC, 1984.

Freedman, Lawrence, and Efraim Karsh, *The Gulf Conflict 1990–1991*, Princeton University Press, Princeton, NJ, 1993.

Fursenko, Aleksandr, and Timothy Naftali, *"One Hell of a Gamble": Khrushchev, Kennedy, and Castro, 1958–1964*, W. W. Norton, New York and London, 1998.

Futrell, Robert Frank, *The United States Air Force in Korea 1950–1953*, Duell, Sloan and Pearce, New York, 1961.

Gaddis, John Lewis, *We Now Know: Rethinking Cold War History*, Oxford University Press, New York, 1997.

GAO (General Accounting Office), *Reserve Forces: Proposals to Expand Call-up Authorities Should Include Numerical Limitations*, U.S. General Accounting Office, Washington, DC, GAO/NSIAD-97-129, April 1997.

George, Alexander, and William E. Simons (eds.), *The Limits of Coercive Diplomacy*, Westview Press, Boulder, CO, 1994 (original date 1971).

Gordon, Michael, and Bernard Trainor, *The Generals' War: The Inside Story of the Conflict in the Gulf*, Little, Brown, Boston, MA, 1994.

Gow, James, "Belgrade and Bosnia—An Assessment of the Yugoslav Military," *Jane's Intelligence Review*, Vol. 5, No. 6, June 1993, pp. 243–246.

Green, Donald P., and Ian Shapiro, *Pathologies of Rational Choice Theory*, Yale University Press, New Haven, CT, 1997.

Haass, Richard, "Sanctioning Madness," *Foreign Affairs*, November/December 1997, pp. 74–85.

Hallin, Daniel, *The Uncensored War*, Oxford University Press, New York, 1986.

Harris, Paul, "Tamil Tigers Intensify War to Establish Homeland," *Jane's International Defense Review*, Vol. 29, May 1996, p. 56.

Hedges, Chris, "Bosnian Serb General Manages to Defy All," *New York Times*, September 14, 1995, p. 10.

Hendrickson, Ryan C., "War Powers, Bosnia, and the 104th Congress," *Political Science Quarterly*, Vol. 113, No. 2, 1998, pp. 241–258.

Hinckley, Barbara, *Less Than Meets the Eye: Foreign Policy Making and the Myth of the Assertive Congress*, University of Chicago Press, Chicago, IL, 1994.

Hiro, Dilip, *Lebanon: Fire and Embers*, St. Martin's Press, New York, 1992.

Hoffman, David, and Ann Devroy, "Too Many Strings Attached, U.S. Finds; Conditions Negate Plan's Possibilities," *Washington Post*, February 22, 1991, p. A1.

Holbrooke, Richard, *To End a War*, Random House, New York, 1998.

Hopf, Ted, *Peripheral Visions: Deterrence Theory and American Foreign Policy in the Third World, 1965–1990*, University of Michigan, Ann Arbor, MI, 1994.

Hosmer, Stephen T., *Psychological Effects of U.S. Air Operations in Four Wars, 1941–1991*, RAND, 1996.

Hunt, Peter C., "Coalition Warfare: Considerations for the Air Component Commander," thesis presented to the School of Advanced Airpower Studies, Air University Press, Maxwell Air Force Base, AL, June 1996.

Huth, Paul K., "Reputations and Deterrence," *Security Studies*, Vol. 7, No. 1, Fall 1997, pp. 72–99.

Huth, Paul, and Bruce Russett, "Testing Deterrence Theory: Rigor Makes a Difference," *World Politics*, July 1990.

Iraqi News Agency broadcasts, January 9–13, 1992, in FBIS NES-92-009, January 14, 1991.

Janis, Irving L., *Groupthink*, Houghton Mifflin Company, Boston, MA, 1982.

Jehl, Douglas, "In NATO Talks, Bosnia Sets Off a Sharp Debate," *New York Times*, January 11, 1994, p. A1.

———, "Only One Arab Nation Endorses U.S. Threat of Attack on Iraq," *New York Times*, February 9, 1998, p. A6.

———, "With Iran's Aid, Guerrillas Gain Against Israelis," *New York Times*, February 26, 1997.

Jentleson, Bruce W., "The Pretty Prudent Public: Post Post-Vietnam American Opinion on the Use of Military Force," *International Studies Quarterly*, Vol. 36, 1992, pp. 49–74.

———, and Rebecca Britton, "Still Pretty Prudent: Post-Cold War American Public Opinion on the Use of Military Force," *Journal of Conflict Resolution*, Vol. 42, No. 4, August 1998, pp. 395–417.

Jervis, Robert, *Perception and Misperception in International Politics*, Princeton University Press, Princeton, NJ, 1976.

Kahn, Herman, *On Escalation: Metaphors and Scenarios*, Frederick A. Praeger, Washington, DC, 1965.

Kahneman, Daniel, and Amos Tversky, "Prospect Theory: An Analysis of Decision Under Risk," *Econometrica* 47, 1979, pp. 263–291.

Kaplan, Fred, *The Wizards of Armageddon*, Stanford University Press, Stanford, CA, 1983.

Keaney, Thomas A., and Eliot A. Cohen, *Gulf War Air Power Survey (GWAPS): A Summary Volume*, U.S. Government Printing Office, Washington, DC, 1993.

Kelleher, Catherine McArdle, "Security in the New Order: Presidents, Polls, and the Use of Force," in Daniel Yankelovich and I. M. Destler (eds.), *Beyond the Beltway: Engaging the Public in U.S. Foreign Policy*, W. W. Norton, New York, 1994.

Khong, Yuen Foong, *Analogies at War*, Princeton University Press, Princeton, NJ, 1992.

Kinzer, Stephen, "Milosevic and Karadzic: 'One Has to Go,'" *International Herald Tribune*, August 12, 1994.

Kirshner, Jonathan, *Currency and Coercion*, Princeton University Press, Princeton, NJ, 1995.

——, "The Microfoundations of Economic Sanctions," *Security Studies*, Vol. 6, No. 3, Spring 1997, pp. 32–64.

Kohan, John, "Fire in the Caucasus," *Time*, December 12, 1994, p. 36.

Kojelis, Linas J., Otto Reich, Ronald Hinckley, and Robert Parry, "Public Diplomacy: Seeking Public Support for Contra Aid Policy," in Richard Sobel (ed.), *Public Opinion in U.S. Foreign Policy: The Controversy Over Contra Aid*, Rowman & Littlefield Publishers, Inc., Lanham, MD, 1993, pp. 151–164.

Lambeth, Benjamin S., "Russia's Air War in Chechnya," *Studies in Conflict & Terrorism*, Vol. 19, No. 4, 1996.

Lancaster, John, "Egypt Urges Diplomacy, Not Force, in U.S.-Iraq Dispute," *Washington Post*, November 14, 1997, p. A35.

Larson, Eric V., *Ends and Means in the Democratic Conversation: Understanding the Role of Casualties in Support for U.S. Military Operations*, RAND, 1996a.

———, *Casualties and Consensus: The Historical Role of Casualties in Domestic Support for U.S. Military Operations*, RAND, 1996b.

———, "Putting Theory to Work: Diagnosing Public Opinion on the U.S. Intervention in Bosnia," in Miroslav Nincic and Joseph Lepgold (eds.), *Being Useful: Policy Relevance and International Relations*, University of Michigan Press, Ann Arbor, MI, forthcoming.

Leurdijk, Dick A., *The United Nations and NATO in Former Yugoslavia*, Netherlands Atlantic Commission, The Hague, Netherlands, 1994.

Levy, Jack, "Prospect Theory, Rational Choice, and International Relations," *International Studies Quarterly*, Vol. 41, 1997, pp. 87–112.

Lieberman, Elli, "What Makes Deterrence Work: Lessons from the Egyptian-Israeli Enduring Rivalry," *Security Studies*, Vol. 4, No. 4, Summer 1995, pp. 833–892.

Lippman, Thomas W., and Barton Gellman, "A Humanitarian Gesture Turns Deadly," *Washington Post*, October 10, 1993, p. A1.

Lorch, Donatella, "Disunity Hampering UN Somalia Effort," *New York Times*, July 12, 1993, p. A8.

Lorenz, F. M., "Law and Anarchy in Somalia," *Parameters*, Vol. 23, No. 4, Winter 1993–1994.

Maoz, Zeev, *National Choices and International Processes*, Cambridge University Press, Cambridge, UK, 1990.

Marr, Phebe, *A Modern History of Iraq*, Westview Press, Boulder, CO, 1985.

Mason, Tony, *Air Power: A Centennial Appraisal*, Brassey's, London, 1994.

Mearsheimer, John J., "The False Promise of International Institutions," *International Security*, Vol. 19, No. 3, Winter 1994/ 1995, pp. 5–49.

Meilinger, Phillip S., *10 Propositions Regarding Air Power*, Air Force History and Museum Program, Air University Press, Maxwell Air Force Base, AL, 1995.

Mierzejewski, Alfred C., *The Collapse of the German War Economy, 1944–1945*, University of North Carolina Press, Chapel Hill, NC, 1988.

Mockaitis, Thomas R., *British Counterinsurgency, 1919–60*, St. Martin's Press, New York, 1990.

Morgan, Patrick M., *Deterrence: A Conceptual Analysis*, Sage Library of Social Science, Beverly Hills, CA, 1977.

———, "Saving Face for the Sake of Deterrence," in Robert Jervis, Richard Ned Lebow, and Janice Gross Stein (eds.), *Psychology & Deterrence*, Johns Hopkins University Press, Baltimore, MD, 1985, pp. 125–152.

Morris, Benny, *Israel's Border Wars, 1949–1956*, Oxford University Press, New York, 1997 (1993).

Mueller, John E., *Policy and Opinion in the Gulf War*, University of Chicago Press, Chicago, IL, 1994.

———, *Quiet Cataclysm: Reflections on the Recent Transformation of World Politics*, HarperCollins, New York, 1995.

Mueller, Karl, "Denial, Punishment, and the Future of Air Power," *Security Studies*, Vol. 7, No. 3, Spring 1998, pp. 182–228.

———, "Strategy, Asymmetric Deterrence, and Accommodation," Ph.D. dissertation, Department of Politics, Princeton University, Princeton, NJ, 1991.

Neff, Donald, *Warriors at Suez*, Simon & Schuster, New York, 1981.

Newport, Frank, "Americans Still Cautious About U.S. Involvement in Bosnia," *The Gallup Poll Monthly*, December 1995, pp. 31–33.

Norton, Augustus Richard, *Amal and the Shi'a*, University of Texas Press, Austin, TX, 1987.

Olson, Mancur Jr., *The Economics of the Wartime Shortage*, Duke University Press, Durham, NC, 1963.

———, "The Economics of Target Selection for the Combined Bomber Offensive," *Royal United Service Institution Journal*, Vol. 107, November 1962, pp. 308–314.

Pape, Robert A., *Bombing to Win*, Cornell University Press, Ithaca, NY, 1996.

———, "The Air Force Strikes Back: A Reply to Barry Watts and John Warden," *Security Studies*, Vol. 7, No. 2, Winter 1997/1998a, pp. 191–214.

———, "The Limits of Precision-Guided Air Power," *Security Studies*, Vol. 7, No. 2, Winter 1997/1998b.

———, "Why Economic Sanctions Do Not Work," *International Security*, Vol. 22, No. 2, Fall 1997, pp. 90–136.

Perlez, Jane, "U.S. Role Is Not to Disarm, Aide to Top Somali Insists," *New York Times*, December 6, 1992, p. 14.

Peterson, Walter J., "Deterrence and Compellance: A Critical Assessment of Conventional Wisdom," *International Studies Quarterly*, Vol. 30, 1986, pp. 269–294.

Pillar, Paul R., "Ending Limited War: The Psychological Dimensions of the Termination Process," in Betty Glad (ed.), *Psychological Dimensions of War*, Sage Publications, Newbury Park, CA, 1990, pp. 252–263.

Pirnie, Bruce R., and William Simons, *Soldiers for Peace: Critical Operational Issues*, RAND, MR-583-OSD, 1996a.

———, *Soldiers for Peace: An Operational Typology*, RAND, MR-582-OSD, 1996b.

Plous, Scott, *The Psychology of Judgment and Decisionmaking*, McGraw-Hill, Inc., New York, 1993.

Pomfret, John, "Bosnian Serb Military Leader Conspicuously Absent from Karadzic Deal," *Washington Post*, July 20, 1996, p. A24.

Psychological Dimensions of War, Betty Glad (ed.), Sage Publications, Newbury Park, CA, 1990.

Ranstorp, Magnus, *Hizb'Allah in Lebanon: The Politics of the Western Hostage Crisis*, St. Martin's Press, New York, 1997.

"Ratko Refuses to Leave the Sinking Ship," *Economist*, September 16, 1995, p. 57.

Richburg, Keith B, "Aideed 'No Longer Part of Process'; UN Officials in Mogadishu Play Down Failure to Arrest Warlord," *Washington Post*, June 19, 1993a, p. A14.

———, "In War on Aideed, UN Battled Itself," *Washington Post*, December 6, 1993b, p. A1.

Riscassi, Robert W., "Principles for Coalition Warfare," *Joint Force Quarterly*, No. 1, Summer 1993.

Robinson, Eugene, "Criticism from Gulf War Allies Strains U.S.-Led Coalition," *Washington Post*, January 20, 1993, p. A25.

Rose, Gideon, "The Exit Strategy Delusion," *Foreign Affairs*, January/February 1998, pp. 56–67.

Sabin, Philip A.G., and Efraim Karsh, "Escalation in the Iran-Iraq War," *Survival*, Vol. 31, No. 3, May/June 1989.

Sagan, Scott D., "From Deterrence to Coercion to War: The Road to Pearl Harbor," in Alexander George and William E. Simons (eds.), *The Limits of Coercive Diplomacy*, Westview Press, Boulder, CO, 1994 (1971), pp. 57–90.

———, "Rules of Engagement," *Security Studies*, Vol. 1, No. 1, Autumn 1991.

Schelling, Thomas, *Arms and Influence*, Yale University Press, New Haven, CT, 1966.

Schiff, Ze'ev, and Ehud Ya'ari, *Israel's Lebanon War*, Simon and Schuster, New York, 1984 (Ina Friedman, translator).

Schmitt, Eric, "Conflict in the Balkans: NATO Commanders Face Grim Choices," *New York Times*, September 14, 1995, p. A1.

Schow, Kenneth C, "Falcons Against the Jihad: Israeli Airpower and Coercive Diplomacy in Southern Lebanon," Air University Press, Maxwell Air Force Base, AL, November, 1995.

Shimshoni, Jonathan, *Israel and Conventional Deterrence: Border Warfare from 1953 to 1970*, Cornell University Press, Ithaca, NY, 1988.

Shogren, Elizabeth, "Rwandans Told World Shares Guilt for Genocide," *Los Angeles Times*, March 26, 1998.

Silber, Laura, "Milosevic Struggles to Pluck Bosnian Thorn," *Financial Times*, August 22, 1994, p. 2.

Slater, Jerome N., "The Dominican Republic, 1961–1965," in Barry M. Blechman and Stephen S. Kaplan (eds.), *Force Without War: U.S. Armed Forces as a Political Instrument*, Brookings Institution, Washington, DC, 1978.

Smith, R. Jeffrey, "Tracking Aideed Hampered by Intelligence Failures," *Washington Post*, October 8, 1993, p. A19.

Smoke, Richard, *War: Controlling Escalation*, Harvard University Press, Cambridge, MA, 1977.

Sobel, Richard, "The Polls—Trends: United States Intervention in Bosnia," *Public Opinion Quarterly*, Vol. 63, No. 62, 1998, pp. 250–278.

Stein, Janice Gross, "The Arab-Israeli War of 1967: Inadvertent War Through Miscalculated Escalation," in Alexander L. George (ed.), *Avoiding War: Problems of Crisis Management*, Westview Press, Boulder, CO, 1991, pp. 126–159.

———, "Deterrence and Compellence in the Gulf, 1990–1991," *International Security*, Vol. 17, No. 2, Fall 1992.

Stieger, Cyrill, "Rifts Among the Serbs," *Swiss Review of World Affairs*, No. 10, October 1994.

Strobel, Warren, *Late-Breaking Foreign Policy*, United States Institute of Peace, Washington, DC, 1997.

Summers, Harry G. Jr., *On Strategy II: A Critical Analysis of the Gulf War*, Dell Books, New York, 1992.

Tanner, Marcus, "Aid Flights Halt on Eve of No-Fly Patrol," *Independent*, April 12, 1993.

Taw, Jennifer Morrison, and Alan Vick, "From Sideshow to Center Stage: The Role of the Army and Air Force in Military Operations Other Than War," in Zalmay M. Khalilzad and David A. Ochmanek. (eds.), *Strategic Appraisal 1997: Strategy and Defense Planning for the 21st Century*, RAND, 1997.

Thies, Wallace J., *When Governments Collide: Coercion and Diplomacy in the Vietnam Conflict, 1964–1968*, University of California Press, Berkeley, CA, 1980.

Tubbs, Maj. James O., "Beyond Gunboat Diplomacy: Forceful Applications of Airpower in Peace Enforcement Operations," Air University Press, Maxwell Air Force Base, AL, September 1997.

USA Today, "Gallup Poll: 56% Support Role in Attacks," April 29, 1999, p. 4A.

USAF (United States Air Force), Air Mobility Command, *Civil Reserve Air Fleet*, Fact Sheet, August 1997.

Vego, Milan, "Federal Army Deployments in Bosnia and Herzegovina," *Jane's Intelligence Review*, Vol. 4, No. 10, October 1992, pp. 445–446.

Venter, Al J., "Hezbollah Defies Onslaught," *Jane's International Defense Review*, Vol. 29, June 1996, pp. 81–82.

Vick, Alan, David Orletsky, John Bordeaux, and David Shlapak, *Enhancing Airpower's Contribution Against Light Infantry Targets*, RAND, MR-697, 1996.

Warden, John III, "Employing Air Power in the 21st Century," in Richard Shultz, Jr., and Robert L. Pfaltzgraff, Jr. (eds.), *The Future of Air Power in the Aftermath of the Gulf War*, Air University Press, Maxwell Air Force Base, AL, 1992, pp. 57–82.

————, "Success in Modern War," *Security Studies*, Vol. 7, No. 2, Winter 1997/1998.

Watts, Barry, "Theory and Evidence in Security Studies," *Security Studies*, Vol. 7, No. 2, Winter 1997/1998.

Waxman, Matthew C., "Coalitions and Limits on Coercive Diplomacy," *Strategic Review*, Vol. 25, No. 1, Winter 1997a.

————, "Emerging Intelligence Challenges," *International Journal of Intelligence and Counterintelligence*, Vol. 10, No. 3, Fall 1997b, pp. 317–331.

Woodward, Susan L., *Balkan Tragedy*, Brookings Institution, Washington, DC, 1995.

Wright, Robin, "Diplomacy: U.S. Officials Concede That Discord Within 29-Nation Alliance Served to Limit Actions Against Iraq," *Los Angeles Times*, January 19, 1993, p. A10.

Yu, Bin, "What China Learned from Its 'Forgotten War' in Korea," *Strategic Review*, Summer 1998 (internet version, accessed August 3, 1998).

Zaller, John, *The Nature and Origins of Mass Opinion*, Cambridge University Press, Cambridge, MA, 1992.